2018

To: Bob & Donna Rochnitzky

SELF and the PHENOMENON of LIFE

with best wishes

[signature]

SELF and the PHENOMENON of LIFE

A biologist examines life from molecules to humanity

RAMON LIM *University of Iowa, USA*

World Scientific

NEW JERSEY · LONDON · SINGAPORE · BEIJING · SHANGHAI · HONG KONG · TAIPEI · CHENNAI · TOKYO

Published by

World Scientific Publishing Co. Pte. Ltd.
5 Toh Tuck Link, Singapore 596224
USA office: 27 Warren Street, Suite 401-402, Hackensack, NJ 07601
UK office: 57 Shelton Street, Covent Garden, London WC2H 9HE

Library of Congress Cataloging-in-Publication Data
Names: Lim, Ramon, author.
Title: Self and the phenomenon of life : a biologist examines life from
   molecules to humanity / Ramon Lim.
Description: Hackensack, New Jersey : World Scientific, 2017. |
   Includes bibliographical references and index.
Identifiers: LCCN 2017000688| ISBN 9789813203778 (hardcover : alk. paper) |
   ISBN 9813203773 (hardcover : alk. paper) | ISBN 9789813203785 (pbk. : alk. paper) |
   ISBN 9813203781 (pbk. : alk. paper)
Subjects: LCSH: Life (Biology)--Philosophy. | Self.
Classification: LCC QH501 .L556 2017 | DDC 570.1--dc23
LC record available at https://lccn.loc.gov/2017000688

British Library Cataloguing-in-Publication Data
A catalogue record for this book is available from the British Library.

Copyright © 2017 by World Scientific Publishing Co. Pte. Ltd.

All rights reserved. This book, or parts thereof, may not be reproduced in any form or by any means, electronic or mechanical, including photocopying, recording or any information storage and retrieval system now known or to be invented, without written permission from the publisher.

For photocopying of material in this volume, please pay a copying fee through the Copyright Clearance Center, Inc., 222 Rosewood Drive, Danvers, MA 01923, USA. In this case permission to photocopy is not required from the publisher.

Printed in Singapore

To my youngest grandchild, Jasper, who at the age
of six confronted me with the question:

"Ye-ye (grandpa), do you believe in God?"

Contents

Preface...ix

1. Introduction: Why *Self*?..................................1
2. An Astronaut's Dilemma.................................9
3. *Self* and the Beginning of Life........................15
4. The Microbial *Self*..79
5. The Plant *Self*...93
6. The Animal *Self*: Molecular Recognition....109
7. The Animal *Self*: Neurobehavioral Correlates............127
8. *Self* and Conscious Experience.....................163
9. *Self* and Emotion..179
10. *Self* and Memory..209
11. *Self* and Free Will...237
12. The Expanded *Self*: Society as *Self*..............257
13. *Self* from Within: The Introspective *Self*.....299
14. *Self*, Realities, and the Transcendents.........315
15. Epilogue: And the Quest Goes On...............335

Appendix A: *Neurotransmitter Structures*......................337
Appendix B: *Organization of the Nervous System*...........339

Appendix C: Relative Anatomical Positions......................341
Appendix D: Approaches to Explore the Brain................343
Glossary..351
Acknowledgments...363
About the Author..365
Index..367

Preface

Ever since I realized the limited span of my own self, I have contemplated what life is about. In an attempt to answer this question, I took up science as a career, as I believe the scientific method is the most reliable way to acquire knowledge. While doing science, I never hesitated to pause for a moment to ponder how my own work, and that of the entire scientific community, add to the understanding of ourselves and the universe.

Philosophy started as a discipline to acquire knowledge and wisdom, and science is an offshoot of philosophy when the subject matter turns to natural phenomena. Ever since the invention of the scientific method, many philosophical issues were solved or rendered irrelevant. But powerful as it is, science has its limitations in the pursuit of global knowledge and wisdom. Where science hits a limit, philosophy steps in. Compared to science, philosophy is intuitive, subjective, and to hard-core scientists, not very reliable. But without it, life is incomplete and unfulfilling, at least to some people including myself.

Some years ago, C. P. Snow pointed out the gap between our two cultures — the sciences and the humanities. I believe the gap cannot be filled by either a purely scientific approach or a purely philosophical approach. A viewpoint that encompasses both is necessary. By standing on the vantage point of "self", defined in this book as a natural system that auto-perpetuates, I hope I have found a common ground between our two cultures. My approach is rigorously scientific, yet in the end it is also philosophical.

Once I retired from a career in scientific research, writing a book on what life is about became my passion. This is not because I have an answer, but because I believe I have enough information to layout an honest perspective of life that includes what is known, what likely will be known, and perhaps what can never be known. Readers who are college graduates with a background in biology and some knowledge in chemistry and a propensity for philosophical issues will most benefit from this book, which is written in a language understandable to both experts and outsiders. Professional jargon, when used, is explained or elaborated in the endnotes, appendices, and glossary.

I would like to thank my three children, Prof. Jennifer Lim-Dunham, M.D., Prof. Wendell Alan Lim, Ph. D., and Caroline Lim Starbird, J.D., for inspiration and encouragement; and my beloved wife of fifty-five years, Prof. Victoria Sy Lim, M.D., for unwavering support and dedication to my lifework and career, frequently at the expense of her own. Their involvement is indispensable in the making of this book.

Lastly, I hope my book will be judged not by how well it sells, but by whether it will be remembered a hundred years hence.

Ramon Lim, M.D., Ph.D.
University of Iowa
Iowa City, IA 52242, USA
August 2016

Chapter 1

Introduction: Why *Self*?

Self is the invariant of life; it is the great divide between the living and the inanimate world.

Overview: *Self is a natural system that acts toward auto-preservation and auto-perpetuation. It is the underlying principle of all living things. Self is the driving force that, under selective pressure, propels evolution forward. As life evolves into the future, the metabolic scheme might change, the genetic mechanism might be different, but self will remain. The approach from self's standpoint opens up multiple perspectives of life, including the observable and the introspective, the physical and the humanistic, and provides a holistic view of what life is about.*

Life is all about *self*.

Like the air we breathe in or the gravity we live with, *self* is with us constantly but hardly noticeable unless we pause for a moment from our busy lives and assume a slightly different vantage point.

In everyday language *self* refers to our own body or our private experience, but in this book I talk about *self* in a much broader sense. I use it to include all forms of life, from the barely living (first life on Earth) to complex plants and animals, including the product of the brain –mind. Living things, big and small, are self-centered. You chase them, and they scurry away; try to kill them, and they run for their lives; when hungry, they look for food, often gobbling up other living things; once fed, they crave sex; after procreation, they guard their young with their lives. Even the most altruistic acts in the biological world, martyrdom included, have a subtle self-directed motivation — preservation of the genome or the clan. What is this *self* that permeates all strata of life?

In this book I define *self* as "*a naturally occurring system whose activity leads to the perpetuation of this same system.*" Let me make clear at the outset that my definition of *self* does not entail consciousness, and thus the term is applicable to all forms of life, including the simplest. This definition, however, does not preclude the entry of *self* into the conscious realm once life arrives at a certain level of complexity, as in higher animals.

Despite scientific progress, how life arose on Earth — or anywhere in the universe — remains largely a mystery. But we do know that life appeared at a critical moment when an assemblage of molecules started to move forward against all obstacles to continue its existence. On the other hand, we cannot predict what life will be like a few billion years hence — it might use a totally different chemical system for harnessing energy and for transmitting information and inheritance. Yet we do know that, if life will still be present then, *self* will be there. *Self* is therefore the first feature of life to appear and the last to go.

I define *self* as a *natural* phenomenon in order to stress its spontaneity, thereby excluding any intelligent design or intervention, human or otherwise. I define *self* as a *system* so as to emphasize the fact that it does not have to be preconditioned on any specific concrete entity, such as molecules like proteins or DNA.[1] This is not to mean that life on Earth is not made of proteins and DNA. Rather, my definition leaves room for the remote possibility that life in some far corners of the universe, if any, may not depend on these same molecules. Furthermore, my "systems" definition emphasizes the importance of interrelationship among parts rather than the individual parts themselves. This abstract definition of *self* has a broad application to many concrete entities of life. At the beginning, when life barely struggled to exist, *self* and life might not be that much different. As time went on, though, life built up in complexity. Nevertheless, just as the edifice of Euclidian geometry can be reduced to a few simple axioms, so too the kaleidoscope of the living world can be understood from the simple concept of "self."[2] As will be seen in subsequent chapters of this volume, almost anything we do or experience, from feeding a baby, going to school, reading a newspaper, playing a violin concerto, shaking hands in the White House, to exploding roadside bombs in Iraq and Afghanistan, and even to launching the space station, can be traced back to that very first feature of the first life that appeared on Earth some 3.8 billion years ago — the *self* and its ceaseless workings to exist *ad infinitum*. Figure 1.1 outlines the relationship between *self* and life, and shows that *self* straddles the two sides of life, the physico-biological and the socio-humanistic. *Self* provides a unified view of life.

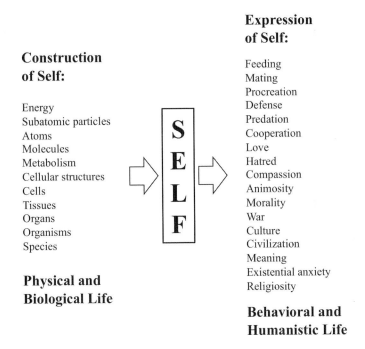

Fig. 1.1. Diagram showing the intersection of *self* and life. The left column depicts the formation of *self* from matter and covers the time span from life's origin to eons of biological evolution, including the renewal of individuals in every generation. Collectively, this is life in the physical and biological sense. The right column depicts what transpires from living things as the expression of *self,* and constitutes life in the behavioral and socio-humanistic sphere. "Self" overlaps both sides of life and serves as a standpoint from which life can be seen as a whole.

You may ask, "Instead of *self*, why not focus on evolution, or genes, or molecules, or DNA, or, in a more abstract context, *replicators*?" My answer is, evolution is not a fundamental principle. Rather, it is the product of interaction between the variation in heritable traits and changing environmental constraints, and hidden behind the two is the *self's* unwavering goal to survive. On the other hand, to define life in terms of molecules is too restrictive, as we do not know whether the molecules of life in remote parts of the universe are the same as those on Earth. The topic of "replicator," especially in relation to genes and DNA, deserves special elaboration and is presented below.

Although the importance of natural selection is indisputable, the substrate of natural selection is elusive and has changed over time. First, it was the organism that was selected. Next, after the discovery of genes and mutation, the genome became the unit of selection. Subsequently, in the 1950s when DNA was discovered to carry the genetic information, DNA and genes were taken to be one and the same, and the focus was shifted to the DNA molecule. It was in this context that the "gene-centered" theory of Richard Dawkins was conceived. DNA was considered to be at the core of life, having the sole function of continuous replication (the *replicator*) and using the other parts of the organism (metabolism, body, brain, etc.) as the *vehicle* to carry out this mission. Evolution then becomes a matter of competition for abundance among genes (or segments of DNA) — a concept called the "selfish gene theory."[3,4]

This "DNA-centered" view of life enjoyed popularity for decades but has gradually lost its appeal (to be elaborated in *Chapter 3: Self and the Beginning of Life*). New evidence shows that evolution is a multidimensional business, and DNA is just one of the dimensions.[5] A new way of thinking places the unit of natural selection back to the organismic level, and considers the organism as a *complex system* in which the activity of the whole is greater than the sum of the parts. DNA is reduced to a tool for the perpetuation of this system. The current book is an offshoot of this line of thinking, and I call this "system of life" *self*. Besides, my definition of "self" permits me to cover both the observable and introspective aspects of life, and allows me to cross the boundary from science to philosophy and humanity.

What about the more abstract idea of a "replicator" as the center of life? In Dawkins' original theory, genes compete by the speed of multiplication, and the faster ones win the race. However, competition by abundance alone is not the theme of life, as unlimited replication does not guarantee long-term preservation. There are natural events that, once started, propagate without an end in sight. Combustion and nuclear chain reaction are prime examples. Absent self-sustainability, they die out as the reactants exhaust. In the animate world, nothing replicates faster than a malignant tumor cell, but without a mechanism for self-regulation or a way to integrate into a larger *self* within the host, the race only leads to its eventual demise — dying heirless as the host perishes. By contrast, all normal living things express a unified "goal" of achieving long-term existence. Replication is just one of the means to achieve it. To say that life can be reduced to a replicator is not so much wrong as inadequate.

The chapters in this book are organized along the evolutionary line. At each stage of evolution, I explain how *self* is constructed and how it is expressed. *Chapter 3* briefly presents the chemistry of the living process and how life, and *self*, emerged from the inanimate world — what we know and what remains to be discovered. In this chapter I stress the simultaneous appearance of *self* and life, since at this primeval stage the two are almost the same. *Chapter 4* discusses how "self" manifests as simple survival instincts in unicellular organisms, such as bacterial defense against invading viruses, and the engulfing of edible particles by an amoeba. I point out in *Chapter 5* how plants, usually taken, mistakenly, as passive and indolent, also express a *self* in the form of defense and communication in times of health and disease. *Chapter 6* discusses how, in animals, immunity distinguishes "self" from "non-self" by strictly molecular recognition, and why your body tries to drive out other people's organs when they are transplanted into you. *Chapter 7* outlines how the development of the nervous system gradually, in evolution time, brings *self* to the conscious level and expresses it in the great varieties of animal behavior. *Chapters 8 to 11* cover some of the important functions of the mind, and show how they are related to *self* — in the latter's enhancement, awareness, and expression. Consciousness (*Chapter 8*) polarizes *self* from the rest of the world, and makes "I" stands out from a crowd. Emotion (*Chapter 9*) provides the driving force for an animal to act for the good of *self*. Memory (*Chapter 10*) provides a sense of continuity to *self* (the biographical *self*) and shapes future actions according to past experience. Free will (*Chapter 11*) strengthens the awareness of *self* by giving it a sense of agency, and enables an animal to make appropriate choices suitable for its own preservation.

In *Chapter 12*, I place *self* in a social context, and show how a group of individuals, when interactive and properly coordinated, can be considered as an expanded *self*. The concept of "expanded self" explains such enigmatic questions as the "goodness" and "badness" of human nature, and why collective mutual destruction of mankind (war) never ends.[6] In *Chapters 13 and 14*, I take a conceptual jump from the observable (the subject matter of science) to the introspective and transcendental (the subject matter of philosophy). This audacious leap may seem illogical. However, once we realize that such deep human concerns as life-and-death, existence-and-nothingness, as well as "meaning", are all extensions of the *self's* craving for its own continuation, the gap may not be as wide as it first appears. The book ends with a positive note that *self*, which initially crystallizes out of the world, eventually longs to be reunited with the wholeness of the universe.

In short, *self* is the common denominator of all living things, in both the biological and spiritual context. Note that I use the terms "spiritual" and "transcendental" in a very broad sense to cover those human activities and mental contents that are not directly needed for everyday survival. They include art, music, literature, drama, etc., in addition to the traditional religious beliefs and rituals. My approach is totally secular, as I do not assume the presence or absence of an immortal soul or a supreme, conscious ruler of the universe.

Other than standing on the vantage point of *self*, I do not see any way to cover so many seemingly disparate facets of life under a single umbrella. This book shows how the simple concept of "self" can explain most, if not all, of what is going on in the biological world. It starts with science but ends up more than science. Hence it can be read either as a scientific treatise with a philosophical overtone, or as a philosophical proposition grounded in science. Facing such a complex subject, I have no choice but to be syncretic, which is far riskier and error-prone than to sit comfortably in one's own niche of specialization. Mindful of this limitation, Erwin Schrödinger, of quantum mechanics fame, in the preface to his seminal book *What Is Life* warned that "I see no other escape from this dilemma than that some of us should venture to embark on a synthesis of facts and theories, albeit with second hand and incomplete knowledge of some of them — and at the risk of making fools of ourselves."[7] Let me take this risk. Life is such an enigma that no attempt at explaining it will be complete and final. I present my view only as an alternative to many others, with shortcomings as well as strengths. Needless to say, the book awaits future corrections and revisions, not only by myself but also by coming generations, as new knowledge arrives.

Notes and References

1. A system is a collection of interrelated, interacting parts that work in a coordinated manner so that the function of the whole is more than those of the individual parts combined.
2. The following illustrates my point. Let me draw a straight line, defined as the shortest distance between two points. The line represents the simplest form of geometry, and it is governed by the axiom as defined. Next, let me draw a three-dimensional structure. The lines and the axiom are embedded in the solid geometry, though they may not be apparent to a casual observer. Likewise, in a complex life form, the principle of *self* (as I define it) may not be obvious but it is always there.
3. Dawkins R. (1976) *The Selfish Gene*. Oxford Univ. Press, Oxford. The title of Dawkins' book implies that genes are independent and individualistic,

each surviving for its own sake, whereas the current view is that genes in an organism cooperate in a highly sophisticated manner in addition to competition. See: Yanai I, Lercher M. (2016) *The Society of Genes.* Harvard Univ. Press, Cambridge, MA.
4. Dawkins R. (1982) *The Extended Phenotype: The Gene as the Unit of Selection.* Oxford: Freeman, p. 114.
5. Jablonka E, Lamb MJ. (2005) *Evolution in Four Dimensions.* MIT Press, Cambridge, MA.
6. My notion of expanded self does have some similarity to Holldobler and Wilson's "superorganism." Nevertheless, their term is mainly applied to the social insects. See: Holldobler B, Wilson EO. (2009) The *Superorganism.* Norton, New York.
7. Schrödinger E. (1944) *What is Life? The Physical Aspects of the Living Cell.* Cambridge Univ. Press, Cambridge, UK.

Chapter 2
An Astronaut's Dilemma

"Is it alive?"

Overview: *Where there is self, there is life.*

Let us start with a trip to the far side of the universe. Imagine you are an astronaut and your spacecraft is approaching a solid planet. You are curious if there is life in this corner of the cosmos. Before departing Earth, you have seen movies with imaginary extraterrestrial beings that came in all shapes and forms, most of which were grotesque and frightening. You probably have read *Barlowe's Guide to Extraterrestrials*, lavishly illustrated with pictures of 51 species, all products of fantasy, of course.[1] You have also read the book *The Black Cloud*, a science fiction novel written in 1957 by the astrophysicist Fred Hoyle, in which an immense cloud approaches the solar system from outer space and comes to rest between the sun and the Earth, threatening eradication of all life by blocking photosynthesis in plants. The cloud is unpredictable in movement and turns out to be conscious, expressing surprise in discovering intelligent life on Earth.[2] While still immersed in fantasy and dazed by confusion, your spacecraft lands, and you are ushered out into the uncharted land.

There you are immediately greeted by three unidentified objects, of which you cannot make heads or tails (see Fig. 2.1). With only bare hands, you have to make a quick judgment as to whether each object is alive. Some questions instantly come to your mind, such as: Does it move? Does it respond to stimulus? Does it increase in size? Does it

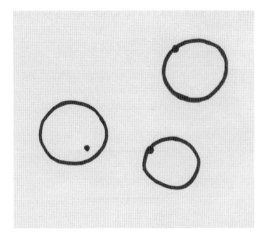

Fig. 2.1. UFO (unidentified foreign objects).

multiply in number? Does it consume energy? Does it put out waste products?

The objects manifest some of the signs of life, but not all, and you still cannot be certain. As you are in no immediate danger of being devoured, you have a few minutes to think. You then reach into your pocket and take out a copy of *Guidebook for Intergalactic Travelers* and look up the definition of life. This is what you find:

(1) Life is an object capable of adaptation by natural selection (biological definition).
(2) Life is a series of chemical reactions carried out in a stepwise, non-equilibrium energy flow (general chemical definition).
(3) Life is a series of carbon-based reactions capable of sustaining itself (organic chemical definition).
(4) Life is a series of enzyme-catalyzed reactions capable of sustaining itself (biochemical definition).
(5) Life is a collection of cells capable of dividing and increasing in number (cell biological definition).
(6) Life is a system of self-replicating macromolecules (molecular-biological definition).

(7) Life is a system capable of mutation and variations in succeeding generations (genetic definition).
(8) Life is a system capable of creating order out of disorder (thermodynamic definition).
(9) Life is a network of information systems (cybernetic definition).

And then there is the "seven-pillar" definition of life, also known as "PICERAS", an acronym for: program, improvisation, compartmentalization, energy, regeneration, adaptability, and seclusion.[3]

There is also a thoughtful definition given by Perret in 1952, which says: "Life is a potentially self-perpetuating system of linked organic reactions, catalyzed stepwise and almost isothermally by complex and specific organic catalysts which are themselves produced by the system."[4]

No doubt, the above definitions and descriptions all fit life observed on Earth. But what about other forms of life that might exist in other parts of the universe? They could be drastically different. For example, they might derive their energy source not from sunlight but from other thermodynamic systems. They might be based on silicon rather than carbon. They might use arsenic rather than phosphorus. They might have a hydrogen-bond rather than covalent-bond backbone. They might thrive in high pressure/temperature, or in very acidic or alkaline conditions. They might have a metabolism based on an organic solvent other than water. They might even exist entirely in solid or gaseous form. Or, they might have an entirely different genetic system.[5]

To accommodate all the possibilities of life, on Earth and elsewhere, here is a general, all-inclusive definition: Life is a self-assembled (spontaneous, growth), self-organized (order, compartmentation, differentiation), self-programmed (information-laden), self-optimized (steady-state internal milieu, homeostasis), self-sustaining (metabolism, maintenance, repair, regeneration), semi-enclosed (plasma membrane), complex molecular system (matter-based), made up of interactive, hierarchical components and networks (cybernetic, auto-regulation), supported energetically by thermodynamic disequilibrium (obeys 2nd law of thermodynamics), carrying out a unified, holistic,[6] self-propelled, self-referential, recursive function (non-repetitive cyclical changes), leading to the perpetuation (preservation, reproduction, adaptation, evolution) of the very same system.

A shorter version would be: Life is a self-assembled auto-regulatory system capable of perpetuating that same system. A much simpler and more common sense definition would be: *Life is a natural occurrence that expresses a self*. Note that the meaning of "self" as expounded in the introductory chapter embodies all the features discussed above.

Figure 2.2 shows a newly hatched chick expressing a strong sense of *self*, while Figure 2.3 portrays a machine seeming to be alive when it really is not. A robot may be programmed to repair itself, to avoid or overcome obstacles, to solve problems, and even to make more robots of its kind, but it lacks *spontaneity*, among other things. It does not mine the metals to make its hardware, nor does it write the initial program to start its fake "life". Worst of all, it does not procure its own energy. As soon as the electric cord is unplugged, it goes *kaput*!

So, simply put, if it has *self*, it is alive.

Fig. 2.2. Strong expression of *self*: hungry chick begging for food. [Permission Bruce Lyon.]

Fig. 2.3. Weak expression of *self*: a robot starved of power. The limited goal-seeking function of a robot does not spontaneously occur; it is a mere extension of that of the programmer. [Permission Bloom County, Cartoonist Group, Berkeley Breathed.]

Notes and References

1. Barlowe WD, Summers I. (1979) *Barlowe's Guide to the Extraterrestrials*. Workman Pub. New York.
2. Hoyle F. (1957) *The Black Cloud*. New American Lib. New York; Easton Press, Norwalk, CT, 1986.
3. Koshland DE Jr. (2002) The seven pillars of life. *Science* **295**: 2215–2216.
4. Perret M. (1952) New Biology. **12**: 68; cited by Luisi PL (1998) About various definitions of life. *Origins of Life and Evolution of the Biosphere* **28**: 613–622.
5. Earth life is based on long chains of carbon atoms forming covalent bonds. As an alternative to carbon, extraterrestrial life may choose to use silicon, which, like carbon, has four electrons in the valence shell and can form four covalent bonds. Chains of silicon atoms 26-member long are known to exist. Arsenic is one row below phosphorus in the periodic table and has bonding properties similar to phosphorus. Arsenic is a potential substitute for phosphorus in life chemistry. Strange forms of life may depend on hydrogen bonds rather than covalent bonds as their backbone, if their habitat happens to have a low temperature (chemical bond strengthens with decreased temperature). Extremophiles are microbes on Earth that thrive on extreme conditions — very high temperature, very high pressure, or very acidic or basic pH. Though unusual to us, they may be the norms of life rather than exceptions in other parts of the universe. Extraterrestrial beings may live in

non-aqueous liquid. Ammonia (NH_3) is in liquid form in an environment having low temperature and high pressure. Liquid ammonia has many properties similar to water. It is a polar molecule and can form hydrogen bonds; it is a solvent for many organic compounds so metabolic processes might take place. Formamide is another solvent similar to water. It is polar and dissolves many things that water dissolves. Furthermore, formamide stabilizes RNA and might provide a favorable environment for an RNA World (see next chapter). Hydrocarbons like methane, ethane, propane, and butane are less polar, but they are abundant in the solar system. They are in liquid form provided the temperature is low enough and the pressure high. In liquid form they are good solvents for organic compounds and might support metabolism. To stretch our imagination further, life might even exist in solid or gas phase. Molecules diffuse very slowly in solids, so imaginary metabolism in solids would take ages, giving solid organisms an extremely long lifespan. In contrast, life in the gaseous form would be too unstable; molecular diffusion would be too fast and chemical reactions too hard to control. Worst of all, gaseous life would not have a confined internal environment, a requirement for life. This is a fatal flaw in the Black Cloud fantasy of Fred Hoyle. For a more thorough discussion of the unusual forms of life that might possibly exist in outer space, see: Benner S. (2009) *Life, the Universe and the Scientific Method.* The FfAME Press, Gainesville, Florida.

6. Holism means the whole is greater than the sum of the parts.

Chapter 3

Self and the Beginning of Life

When a lump of matter expresses a "goal," self, and therefore life, is formed.

Overview: *Although simple organic molecules are abundant in the universe, the origin of life is mired in controversy. How life started is both a problem of chemistry and of organization. Three components have to be simultaneously present: (1) a privileged internal environment for efficient metabolism and discrete information transfer; (2) an enzyme-based metabolism that harnesses energy for building an enclosed environment, and for making information-carrying molecules; (3) a mechanism for information transfer from one generation to the next and from the genes to enzymes. Life emerged when all the three conditions were met. The birth of life was likely to be arduous and tumultuous, but once formed, life propelled itself along a course of no return. The birth of life is also the birth of self. Self, defined as a natural system that seeks its own perpetuation, is the driving force of evolution. It is the axis around which life evolves in response to natural selection.*

"Where do we come from? What are we? Where are we going?" These are the words inscribed by Paul Gauguin in his famous Tahiti painting, now hung in Boston's Museum of Fine Arts. Indeed, finding out how we originated is always daunting and tantalizing. Each religion has its own story of how humans were created, or how life comes from reincarnations of endless series of previous lives. Mythologies abound in tribal societies regarding the origin of mankind.

In everyday life, we bump into living things of some sort whenever we move around. Life is so prevalent and abundant on Earth that, until the nineteenth century, people thought living things pop out of nothing, given the right condition, such as maggots appearing from spoiled food. The debate raged on until 1859 when Louis Pasteur settled the problem of spontaneous generation. Pasteur's famous experiment consisted of keeping boiled broth in a flask having an "S" shaped neck. The broth did not get spoiled despite the free flow of room air into it. However, the soup quickly became cloudy if the flask neck was broken or if the flask was tilted to let the broth touch the swan-neck. (Apparently the neck served as a trap preventing microbes entering from the outside air.) The conclusion was clear: Life comes only from life.

Today, science tells us that more complex lives came from simpler lives, and in this manner extrapolates to the simplest, primordial unicellular life. But how did the very first life start? When in the history of Earth did inanimate matter first turn into living matter? The clue is chemistry. Here, a brief review of chemical concepts is necessary and will be helpful. (For those who are not comfortable with chemistry, feel free to skip the first two thirds of this chapter and go to the section "*What have we learned about the origin of life?*")

3.1 How Chemistry Shapes the World

Everything we come in contact with is composed of molecules, living things not excepted. Molecules are formed from atoms by chemical bonds. Atoms, in turn, are made up of three smaller components: protons, neutrons and electrons, the first two being made up of quarks (three quarks for each proton and neutron, out of six possible choices). The varied combinations of protons, neutrons and electrons produce all the elements of matter. Protons and neutrons are approximately of equal mass, but electrons are much smaller, about 1/2000 that of a proton. However, it is the latter that imparts distinctive chemical properties to each element. While protons and neutrons pack tightly inside the nucleus, the tiny electrons swirl around at an incredible speed in the almost empty space surrounding the nucleus (in quantum mechanics, electrons are visualized as clouds). The electrons and protons are equal in number, producing electrical neutrality, whereas neutrons vary in number even for the same element. Elements having different number of neutrons are called isotopes.

All chemical compounds are formed by interaction of electrons, which are grouped by energy levels into electron shells, from the lowest level (innermost shell) to the highest (outermost shell). *Pauli's exclusion principle* dictates that no two electrons in the same atom can simultaneously occupy the same quantum state, denoted by a set of four quantum numbers.[1,2] This ordering rule demands that each electron finds its "niche" in the shells surrounding the nucleus. Those in the outermost orbit, called valence electrons, constantly "rub shoulders" with neighboring atoms and negotiate by giving away, taking in, or sharing electrons, in effect creating chemical bonds to form molecules.[3] Simply put, outer shell electrons determine what molecules can be found in the universe. Out of the four fundamental forces of nature (strong and weak interactions, electromagnetism, and gravity), only electromagnetic force, being responsible for chemical bond formation, plays a direct and major role in life.[4] Thanks to the electrons, a kaleidoscopic world unfolds before our eyes and gives us everything we encounter — dandelions, bacteria, butterflies, babies, and even you and me.

Out of the 92 natural elements present in the universe, only a handful is needed to form life. The major ones are (in decreasing order of abundance) oxygen, carbon, hydrogen, nitrogen, calcium, phosphorus, potassium, sulfur, sodium, chlorine, magnesium, and traces of metals. The first four are of lower atomic weight, located in the first two rows of the periodic table, and make up 96% by weight of a living body (Table 3.1). Hydrogen is interesting in that it is the lightest element, consisting of only one proton and one electron, and was the first to form in the beginning of the universe 13.7 billion years ago. All other elements of higher atomic weight up to iron were latecomers, being products of nuclear fusion taking place in stars. Elements heavier than iron were results of supernova explosion.[5] In the present state of the visible universe, hydrogen makes up 80% of matter, most of the remaining 20% being helium.

Table 3.1. Major Elements in the Human Body

Element	Atomic number	% Total body weight	% Total body atoms
Hydrogen	1	10	63.0
Oxygen	8	65	23.5
Carbon	6	18	9.5
Nitrogen	7	3.0	1.4
Sum		96	97.4

It is interesting to note that the most abundant elements in our body (Table 3.1) are readily found not only on Earth, but also in the stars and interstellar space, evidence that a chemical continuum exists between the animate and the inanimate worlds.

3.2 How Chemistry Shapes Life

Molecules in living things are much more complex than those in the inanimate world. Life on Earth is based on carbons connected in long chains. Carbon forms relatively stable covalent bonds not only with another carbon (C-C), but also with hydrogen (C-H), oxygen (C-O), nitrogen (C-N) and, less frequently, sulfur (C-S). Phosphorus is frequently found in biological molecules, but it is never directly connected to carbon; it does this through oxygen in the form of phosphoric acid. Molecules with two or more phosphates connected in tandem (called pyrophosphates) contain high levels of energy, such as ATP, an abbreviation for adenosine triphosphate. In fact, ATP is the energy currency that is used to synthesize almost all the components inside a cell. It makes possible the incorporation of molecules into the living system. Table 3.2 lists the major types of compounds found in the living system.

What, then, is in the chemistry of life? Let me start with a cell (Fig. 3.1). First and foremost, each cell is enclosed with a membrane, called plasma membrane, made up of two layers of phospholipids stacked back to back, with specialized proteins studded here and there to control the entry and exit of molecules and ions (charged molecules

Table 3.2. **Types of Compounds in Living Matter**

1. **Amino acids**: Building blocks of proteins; the latter comprises structural proteins, membrane components, enzymes, hormones, signal transducers, transcription factors, growth factors, antibodies, etc. Proteins are built up of 20 different kinds of amino acids in various combinations and permutations, all arranged in a linear sequence.
2. **Nucleobases**: Two types of purines (adenine, guanine) and three types of pyrimidines (uracil, thymine, cytosine); as components of DNA and RNA, as coenzymes, as high-energy phosphate compounds. A nucleobase and a sugar make up a nucleoside; a nucleoside plus a phosphate make up a nucleotide. In DNA the sugar is deoxyribose and the bases are adenine (A), guanine (G), thymine (T) and cytosine (C). In RNA the sugar is ribose and the bases are adenine (A), guanine (G), uracil (U) and cytosine (C). In DNA the polymers of nucleotides form two complementary chains (double helix), with the nucleobases from one chain paired with those of the other chain in the following manner: A-T; G-C. In RNA, T is replaced by U. Because the RNA chains are flexible, the three dimensional structure of RNA is much more complex than double-stranded DNA; depending on the base sequence, RNA may form various degrees of intra-chain complementary pairing.
3. **Carbohydrates**: As substrates of metabolism; as a form of energy storage; as protein/lipid modifier; as components of connective tissue, cell wall, and nucleic acids; as structural materials in plants.
4. **Lipids**: As backbones of cell membranes and cell wall; as a form of energy storage; as hormones and signal transducers.
5. **Vitamins and coenzymes**: As small molecules that assist enzyme action.
6. **Others**: Water, electrolytes (sodium, potassium, calcium, magnesium, chloride), phosphoric acid and its derivatives (including the "high-energy" ATP), carbon dioxide, hormones, neurotransmitters, second messengers, minerals and trace metals, waste products, and other minor components.

Note: Even the simplest bacterium today contains over 5,000 different organic compounds.

or atoms), and to serve as sensors of outside environment (receptors). In bacteria and plants, the plasma membrane is protected outside by a cell wall, made of lipids and carbohydrates. Inside the cell there is a long thread of DNA (deoxyribonucleic acid), the carrier of the genetic message (Fig. 3.2). In plants and animals, DNA is bound to a group of small proteins called histones and neatly packaged into a compact "ball" (chromatin) inside a membranous enclosure — the nucleus. When a cell is ready to divide, the chromatin is repackaged into rod-

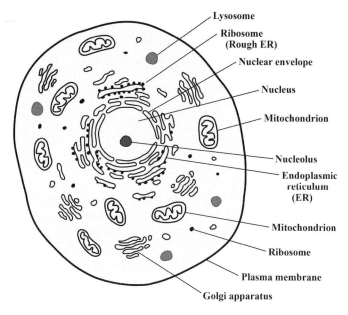

Fig. 3.1. Drawing of a typical animal cell showing its major components. The entire cell is enclosed by a membrane called the plasma membrane. The most prominent structure inside the cell is a nucleus containing the genetic material (DNA) in the form of chromatin. In the nucleus, the nucleolus is the part of chromatin that is dedicated to the synthesis of ribosomes. The nucleus is surrounded by a membrane with openings permitting the entry and exit of materials. Inside the nucleus, messenger RNAs are synthesized according to the information in the DNA. In the cytoplasm, proteins are synthesized on ribosomes according to the information in the messenger RNA. Endoplasmic reticulum (ER) is the membrane in the cytoplasm that is continuous with the nuclear membrane. Ribosomes engaged in protein synthesis are adhered to the ER, forming "rough ER". Nascent proteins destined for export are channeled through a stack of smooth ER called Golgi apparatus. Also prominent in the cytoplasm are the football-like structures called mitochondria, which are responsible for harnessing energy (in the form of ATP) from food. Lysosomes contain digestive enzymes for breaking down cellular wastes. Other structures inside the cell, including peroxisomes, actin fibers, microtubules, and centrosome, are not shown.

like structures called chromosomes. (In bacteria, there is no distinct nucleus, so the DNA is loosely packed in a structure called a "nucleoid." Further, bacterial DNA is distinct from higher forms of DNA in that it has a circular form.) Present in the cytoplasm of all cells are tiny protein-making factories called ribosomes, which are composites

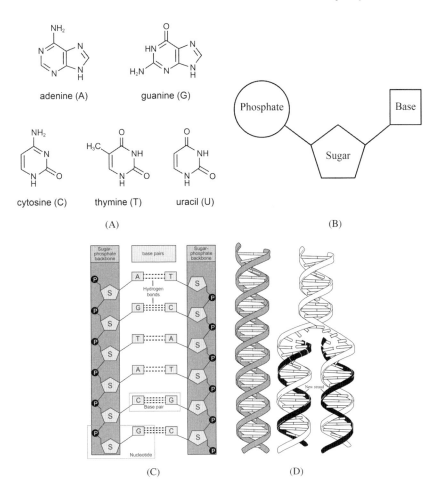

Fig. 3.2. Structure of DNA. (A) The nucleobases in DNA consist of **adenine (A), guanine (G), cytosine (C), and thymine (T)**. The base pairing between two strands of DNA is **A-T** and **G-C**. (The bases in RNA differ by the fact that thymine is replaced by uracil, which forms **A-U** pairing.) (B) A nucleotide is composed of a nucleobase (rectangle) combined with a 5-carbon sugar (pentagon) which in turn is joined to a phosphate group (circle). In DNA, the sugar is deoxyribose, whereas in RNA it is ribose. A nucleotide minus the phosphate group is called nucleoside. (C) Two complementary chains of DNA joined by the hydrogen bonds of **A-T** and **G-C** pairing. Each individual chain is formed by alternating the sugar (S) and phosphate (P) moieties. The right hand side shows a 3-dimensional structure of DNA, with the two strands coiled up to form a double helix, in which the complementary nucleobases form the rungs of a twisted ladder, and the phosphate-sugar backbones form the two side-rails. (In contrast, RNA chains are flexible and do not form a double helix.) (D) Drawing showing how DNA replicates by unwinding the double helix, followed by the building of two new complementary strands. [US Natl. Lib. Med.]

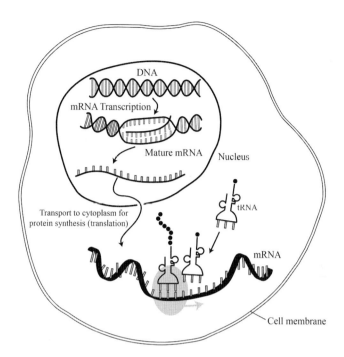

Fig. 3.3. Schematic of a cell showing the transcription of messenger RNA (mRNA) from DNA in the nucleus, the transport of mRNA to the cytoplasm, and the translation of mRNA into protein on a ribosome (shaded round structure). Transfer RNAs (tRNAs) are shown carrying amino acids to match the nucleotide sequence on the mRNA. During translation, the protein chain grows while attached to the tRNA. [US Natl. Lib. Med.]

of RNA (ribonucleic acid) and proteins. Genetic messages encoded in the DNA are *transcribed* into messenger RNA (mRNA) in the nucleus and transported to the ribosomes in the cytoplasm, where the nucleotide (or nucleobase) messages (ATGC in DNA, and AUGC in mRNA) are *translated* into the amino acid sequence of a protein (Fig. 3.3). The detail of translation is depicted in Fig. 3.4. Translation is accomplished by a cooperative action of mRNA and transfer RNA (tRNA). As mRNA lines up on the ribosome, each tRNA carries a specific amino acid matching the genetic code in the mRNA (Table 3.3). The unit of genetic code is a triplet of nucleotides called *codon*. There are twenty kinds of amino acids in a protein, but there are more than sixty codons

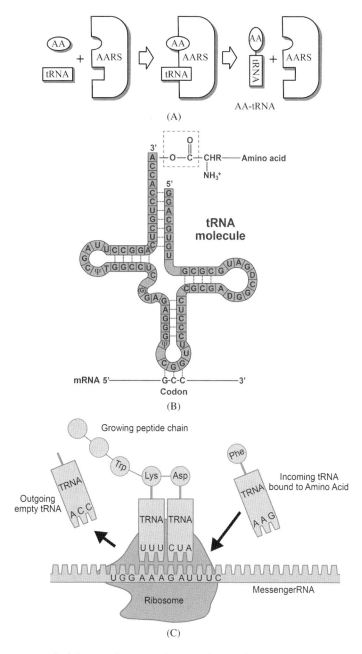

Fig. 3.4. Detail of the translation machinery. The machine converts nucleic acid language into protein language according to the genetic code. (A) Schematic drawing showing how the enzyme aminoacyl-tRNA synthetase (AARS) matches the amino acid (AA)

in the mRNA. Nature resolves this discrepancy by assigning more than one codon to some amino acids. This redundancy of the codes is called *degeneracy*. All tRNAs possess a specific codon-recognition site called *anticodon*, and there are over sixty tRNAs corresponding to the number of codons in the mRNA. The job of matching the twenty amino acids to so many tRNAs is the work of a set of remarkable enzymes called aminoacyl-tRNA synthetases (AARS). There is one AARS assigned to each amino acid. Each AARS, therefore, recognizes its specific amino acid on one part of its molecule and at the same time the corresponding tRNA on another part. What is remarkable is that, in the event of degeneracy, one AARS is able to match one amino acid to several tRNAs. This versatile matching ability has strong implication in the origin of life and will be discussed later.

In summary, the direction of information flow for the biopolymers is DNA → mRNA → protein, a phenomenon known as the Central Dogma (Fig. 3.5). The other function of DNA, which is to replicate itself by complementary base pairing, is depicted in Fig. 3.2.

Proteins are polymers of amino acids linked together by peptide bonds. Out of a great variety of amino acids found in nature, only twenty kinds are assembled into a protein. Just as the 26 letters of the English alphabet can be arranged in limitless combinations and permutations to make words, the living system uses the 20 amino acid letters to construct thousands of different proteins. A typical protein chain contains from about 100 to several thousand amino acids, the sequence of which decides what 3-dimensional appearance (conformation) the protein molecule assumes. Conformation is extremely important for protein

Fig. 3.4. (*Continued*) to its cognate tRNA. The enzyme catalyzes the formation of a covalent bond between an amino acid and the tRNA, now called aminoacyl tRNA (AA-tRNA). (B) Detail of an aminoacyl-tRNA showing an amino acid linked to the adenylic acid nucleotide at the 3′ end (on top) of the tRNA. The location of the anticodon, which is to match the complementary sequence (codon) on the mRNA, is shown at the bottom. (C) Drawing showing the assemblage of a protein chain of amino acids on a ribosome by matching the codons (on messenger RNA) and anticodons (on tRNA). [B and C: US Natl. Lib. Med.]

Table 3.3. The Standard Genetic Code

The code is composed of three-letter words (codons) on the messenger RNA, each codon corresponding to one amino acid. Each letter has four possibilities: U (uracil), C (cytosine), A (adenine), G (guanine). There are 64 codons, but only 20 amino-acid building blocks available for making proteins, resulting in some amino acids being assigned to more than one codon. For example, methionine and tryptophan are each assigned only one codon; leucine, arginine, and serine are each assigned six codons; while others have an intermediate number of codons. Some codons (UAA, UAG, UGA) do not code for amino acids but signify the end of a protein (stop codon).

Amino Acids	Codons
Alanine	GCA GCC GCG GCU
Arginine	AGA AGG CGA CGC CGG CGU
Asparagine	AAC AAU
Aspartic Acid	GAC GAU
Cysteine	UGC UGU
Glutamic Acid	GAA GAG
Glutamine	CAA CAG
Glycine	GGA GGC GGG GGU
Histidine	CAC CAU
Isoleucine	AUA AUC AUU
Leucine	UUA UUG CUA CUC CUG CUU
Lysine	AAA AAG
Methionine	AUG
Phenylalanine	UUC UUU
Proline	CCA CCC CCG CCU
Serine	AGC AGU UCA UCC UCG UCU
Threonine	ACA ACC ACG ACU
Tryptophan	UGG
Tyrosine	UAC UAU
Valine	GUA GUC GUG GUU

Fig. 3.5. The Central Dogma of molecular biology as proposed by Francis Crick. The flow of information is from DNA to RNA and then to protein. The flow of information from DNA to DNA happens only during cell replication.

function, as a protein that loses its native conformation loses its biological activity (see Fig. 3.6). There are over 10,000 kinds of proteins in a mammalian cell, and about half of these are enzymes. Enzymes are the workhorse molecules. They perform a function called catalysis, which speeds up a chemical reaction without itself being altered. All molecules change from one form to another according to thermodynamics, i.e., from one with a higher energy level to one with lower. Enzymes accelerate this reaction by decreasing the energy barrier needed for the change. In the living system chemical reactions take place as quickly as milliseconds. Without enzymes these changes may take months or years, too slow to cope with the immediate needs of life. Hence, almost all biochemical reactions are aided by enzymes. These include the intake of nutrients, digestion of food, harnessing of energy, DNA replication, transcription, translation, and even the synthesis of enzymes themselves. The storage of energy in the form of carbohydrate and fat, and the synthesis of components of the physical structures of a cell are all the work of enzymes.

Some proteins are important for signal transduction, i.e., for transmitting a message from one part of the cell to another, such as from the cell surface to the nucleus in a stepwise manner.[6] Other proteins are regulators of genes. One type, called a transcription factor, activates a gene and leads to the synthesis of a protein; the opposite type, called a repressor, blocks the activation. Some proteins serve as cytoskeletons, providing rigidity and motility to the cells. Others are constituents of the physical framework of a cell, usually in combination with carbohydrates and lipids. The different types of proteins are shown in Table 3.4.

In addition to the nucleus, there are other visible structures inside a cell, collectively designated as the organelles. Mitochondria are the power plants where carbohydrates are oxidized (called "respiration") with O_2 to make ATP (adenosine triphosphate), the energy currency of a cell. The harnessing of energy from carbohydrates goes through four stages. The first stage is the glycolytic pathway in the cytoplasm in which a glucose molecule is turned into pyruvate. The second is the tricar-

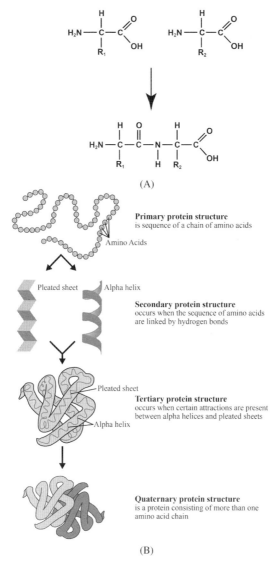

Fig. 3.6. Formation of protein and its conformation. (A) Two individual amino acids joined by a peptide linkage (amide bond) to form a dipeptide. "R" refers to the side chain of an amino acid. The twenty different amino acids in a protein differ from one another by the side chains they carry. Therefore, there are twenty different kinds of side chains in a protein molecule, the combined effect of which imparts the uniqueness of a protein. (B) Drawing showing how a protein (a long chain of amino acids) is folded to form 3-dimensional secondary, tertiary, and quaternary structures. [US Natl. Lib. Med.]

Table 3.4. Types of Proteins

(1) **Structural proteins**
Building materials for cellular components in an aqueous environment, including plasma membrane and intracellular membranes (e.g., nuclear membrane, endoplasmic reticulum, Golgi apparatus) and organelles (e.g., mitochondria, lysosomes); usually in combination with lipids and carbohydrates.

(2) **Functional proteins**
 (a) **Catalytic proteins**– Also known as enzymes. Expedite the making and breaking of chemical bonds; synthesize and degrade biomolecules such as proteins, DNA, RNA, lipids, and carbohydrates; help in harnessing energy.
 (b) **Regulatory proteins**– Monitor the process of cell division; involved in myriads of functions including gene regulation, signal transduction, transcription, translation, protein folding, protein translocation, protein disposal, intracellular homeostasis, intercellular communication, and interaction with environment; movement at all levels; defense against invaders (immunity) and noxious chemicals; transport of oxygen, nutrients, and minerals.

boxylic acid (TCA) cycle taking place inside the mitochondria, and the third stage is the electron transport chain (also in mitochondria) where electrons and protons (hydrogen) are separately carried along the mitochondrial membrane. The final stage is called oxidative phosphorylation, in which hydrogen meets oxygen (taken in from the air) to form water, providing energy to generate ATP. The general chemical equation for the oxidation of glucose is $C_6H_{12}O_6 + 6O_2 \rightarrow 6CO_2 + 6H_2O$. Overall, 38 molecules of ATP are generated from one molecule of glucose. Figure 3.7 is a simplified diagram of energy utilization starting from glucose. Figure 3.8 shows the complex network of overall metabolism including energy utilization and the biosynthetic pathways.

Other organelles are lysosomes, which are the "garbage disposal" of the cell where waste molecules are chewed up with hydrolytic enzymes; and peroxisomes, which contain enzymes capable of degrading reactive oxygen species (toxic byproducts of oxidative phosphorylation). The endoplasmic reticulum (ER) is a series of membranous channels within the cytoplasm on which the ribosomes (see Figs. 3.1, 3.3, and 3.4C) are anchored. After proteins are synthesized on the ribosomes, they are transported through the endoplasmic reticulum to different parts of a

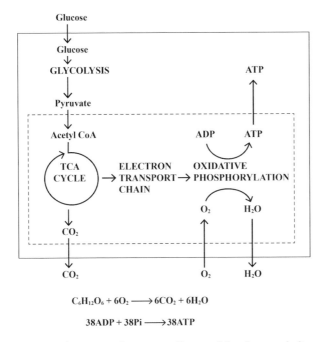

Fig. 3.7. Diagram of energy utilization as illustrated by the metabolism of glucose. The solid line enclosure represents a cell; the dotted line, a mitochondrion inside the cell. One molecule of glucose (a 6-carbon molecule) that we take in from food enters the cell and goes through a process called glycolysis (9 steps), ending up with 2 molecules of pyruvate (3-carbon molecule). Pyruvate enters the mitochondrion where it is converted to acetyl CoA after shedding 1 carbon (as CO_2). The acetyl group then enters the TCA cycle. One cycle of TCA (8 steps per cycle) converts the 2-carbon acetyl group into 2 CO_2, while sending the electrons through the electron transport chain. During electron transport, a proton gradient (created by hydrogen atoms missing the electron) is produced across the inner membrane of the mitochondrion, providing energy for the conversion of ADP (adenosine diphosphate) to ATP (adenosine triphosphate), coupled with the formation of water by the combination of oxygen and hydrogen. The last step is called oxidative phosphorylation. The overall event consists of the reaction of glucose ($C_6H_{12}O_6$) and oxygen to form carbon dioxide and water, with the resulting formation of ATP from ADP and inorganic phosphate (Pi). Starting from one molecule of glucose, a total of 38 molecules of ATP are produced (2 from glycolysis and 36 from reactions inside the mitochondrion).

cell to be utilized. It is in the ER where nascent proteins are folded into the correct shape (there are special proteins that assist in this process) and necessary additions (such as carbohydrates) are installed. It also serves as a "post-office" where proteins are chemically tagged with

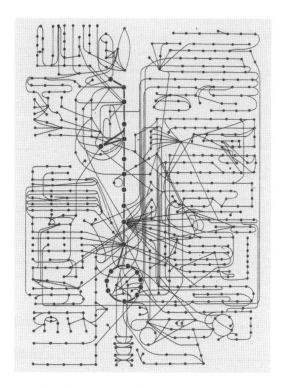

Fig. 3.8. Diagram of metabolic network in the human cell, consisting of more than 500 enzyme-catalyzed chemical reactions. The dots represent metabolites, while the lines, steps of chemical transformation, either for energy production or biosynthesis. [See Note 68; permission Oxford Univ. Press.]

"addresses" of their destinations inside and outside (for secretion) the cell. The ER is also the place where lipids and complex carbohydrates are synthesized. The Golgi apparatus is a modified endoplasmic reticulum, appearing like a stack of membranes, where proteins are packaged for secretion. (See Fig. 3.1.)

Chloroplasts are large organelles found in photosynthetic plants. The chlorophyll inside the chloroplasts absorbs solar energy in the form of light and uses it to convert carbon dioxide and water to carbohydrate and oxygen. Photosynthesis is a reductive process for carbon, the opposite of oxidative process in mitochondrial respiration. The general equation is also the reverse of sugar oxidation: $6H_2O + 6CO_2 \rightarrow C_6H_{12}O_6 + 6O_2$. The sugars produced by photosynthesis are usually stored in the form of polysaccharides such as starch and cellulose. Please note that, in plants, chloroplasts and mitochondria coexist in the same cell, as the latter is needed for energy utilization.

What is as important as the visible structures is the internal environment of the cell. Biochemical reactions take place in an aqueous solution under a narrow range of conditions. These include the total salt concentration (ionic strength), the hydrogen concentration (pH or acidity), the osmotic pressure (concentration of total particles), the temperature, and myriad other nuances. Any deviation from the optimal and enzymes will cease functioning and the life process will be derailed. A complex network of feedbacks and regulations is built in to safeguard a constant and stable internal environment — not a state of equilibrium but a steady state.[7] The overall mechanism leading to a constant internal environment is called homeostasis.

There is also the requirement of energy supply, which in extant life comes from the following sources: (1) electromagnetic waves (sunlight) for photosynthetic organisms; (2) thermal energy from the Earth; (3) exothermal energy from oxidation-reduction of inorganic molecules; (4) intake of organic molecules, especially carbon-rich compounds; (5) engulfment of other organisms (a shortcut to recharge).

Energy is needed to overcome the second law of thermodynamics, which mandates that in a closed system the natural tendency is to change from an orderly to a disorderly situation. Schrödinger was the first to point out this paradox in living things, as the self-organization of life is in apparent violation of the law.[8] This paradox is resolved once we recognize that an organism is neither a closed nor a chemically homogenous system. The hierarchical nature of life makes it possible to superimpose order at a higher level on the disordering tendency of matter at a lower level, at the expense of energy extracted from the surrounding to construct the discrete structure of a living cell.

In summary, a cell, even the simplest one, is not a bag of homogeneous protoplasm, but a miniature metropolis complete with a city hall, residential buildings, commercial establishments, factories, thoroughfares, bridges, subways, sanitary systems, post office, police force, electric power supply, and a conglomeration of numerous other things.

3.3 Kant's Insight Applied on the Chemistry of Life

The above description is a very sketchy outline of the goings-on in a cell. Current knowledge on the chemistry of life can fill up many volumes, and more is still coming, but the details are unnecessary for the purpose of this book. Suffice it to say that the complexity of life is far more than anything else found in the universe, and this complexity is dynamic, with numerous intersecting cycles and networks (Fig. 3.8), the sum of which results in the preservation of the whole — the cell, or the entire organism in the case of multi-cellular life. Over two hundred years ago, Immanuel Kant had the insight that an organism is "that in which everything is both a means and an end."[9] Certainly, Kant made the statement without the benefit of today's chemical knowledge. But no matter, the conclusion remains the same: in life, all processes (the means) lead to the maintenance of the whole (the end), which itself is but a collection of the same processes (the means) that leads to the next phase of the whole. In other words, the life process is circular and recursive, which in the context of Darwinian evolution becomes spiral.

The recursive nature of the living process is the outcome of three interdependent components (Fig. 3.9): (1) A physical *enclosure* that provides a privileged environment, protecting the cell from the ravages of the outside world, yet permitting the selective exchange of nutrients (inward) and waste products (outward). This internal environment sets an optimal condition for metabolism and information transfer. (2) An enzyme-based *metabolism* for harnessing energy and for making all parts of the cell, including the membrane enclosures, the enzymes themselves, and the information-carrying molecules. (3) A *genetic* mechanism for information transfer from genes to proteins (protein synthesis) and from genes to genes (replication). Note that all the three components must be simultaneously functioning for the living process to continue.

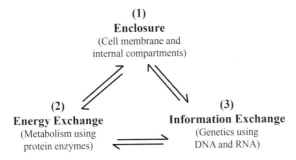

Fig. 3.9. Tripartite interdependent components of a living system: (1) an enclosed environment consisting of cell membrane and internal compartments; (2) energy exchange consisting of metabolism utilizing protein enzymes; (3) information transfer from DNA to DNA (replication) and from DNA to RNA and to protein. Note that (1) is needed for (2) and (3); (2) is needed for (1) and (3); (3) is needed for (1) and (2).

3.4 The Uniformity of Biochemistry

It is puzzling that, despite the great diversity of life forms, the underlying chemical processes are so astoundingly similar. I myself may look different from a cockroach that infests my kitchen sink, but we both use the same metabolic pathways, the same ATP as energy source, and the same DNA for inheritance. The cockroach and I both survive on the same oxygen molecules and exhale carbon dioxide. The lettuce I eat for salad differs from me mainly in its ability to convert electromagnetic energy from sunlight to chemical energy, a fact that I take advantage of by ingesting it as food. True, there are microbes (the anaerobes) that abhor oxygen and can only thrive in a reduced environment — a vestige of past life before oxygen appeared on Earth — but these are minor variations that do not overthrow the general outline of uniformity.

What is most amazing is the near-universality of the genetic code for translating DNA (and RNA) language into protein language (to be elaborated later). Though seemingly randomly chosen at first sight, the three-letter codes are identical not only across species but also across major branches of life, such as between plants and animals.[10]

The biochemical uniformity poses some perplexing questions. Did life occur only once on Earth from a single organism or from a closely related interbreeding population (a unique, contingent event)? Or did it happen multiple times as a result of thermodynamic imperative?

3.5 The Question of Life's Origin

The Earth is estimated to be 4.6 billion years old, and there is abundant fossil evidence that life existed 3.8 billion years ago.[11] Since at the start the Earth was in a chaotic state hostile to life, being constantly bombarded by meteorites, comets and other objects, a reasonable guess is that the first life appeared between 4.2 and 3.8 billion years ago.

The origin of life is a unique historical event that cannot be faithfully reenacted. The best we can do is to obtain observable facts or experimental clues that might help us piece together a plausible hypothesis. How life started is both a chemical and organizational problem. Although the precursors are everywhere, the problem is how they came together to form an intricate network. The three interdependent components outlined above had to be ready at the same time for the wheel of life to start turning and continue in perpetuation.

Ever since the original "primordial soup" idea suggested by Darwin,[12] there has been no shortage of theories on the origin of life, among which are: deep-sea hydrothermal vents; underground aquifers; partially frozen lakes; comets; clay surface… In this chapter I shall deal with some of the proposed scenarios and stress the pertinence of *self* to the issue of life's origin.

3.6 Ingredients of Life are Everywhere in the Universe

Simple organic molecules that can potentially serve as starting materials for life have been detected in meteorites, comets, cosmic clouds, and interstellar dust. Meteorites and comets are leftovers from the birth of the solar system; they are molecular fossils preserved in time and therefore are likely to be present in the early Earth. Cosmic clouds and dust, believed to be leftover of supernova explosions, are cradles of new galaxies and stars, including our sun.

In 1969, a meteorite made up of carbonaceous chondrites landed in Murchison, Victoria, in Australia. Carbonaceous chondrites originated from matter that existed in the planet-forming gas of the early solar system. This is one of the most studied space rocks because it is unequivocally recorded and is free of contamination. A total of 14,000 specific organic compounds, including more than 70 amino acids (such as glycine, alanine, and glutamic acid), have been identified.[13] Polyols, the polyhydroxylated compounds including sugars and other carbohydrates that can form backbones of nucleic acids and components of cell membrane, are indigenous to the meteorite.[14] In addition, xanthine (a precursor of guanine) and uracil (a pyrimidine component of RNA) have been detected.[15] See Table 3.5 for comparative abundance of organic compounds in the Murchison meteorite.

Table 3.5. Organic Compounds in Murchison Meteorite

Amino acids	++
Carboxylic acids	+++
Dicarboxylic acids	++
Hydroxy acids	++
Sulfonic acids	++++
Basic N-heterocyclics	+
Purines & pyrimidines	+
Pyridine carboxylic acids	+
Amides	++
Amines	++
Alcohols	++
Aldehydes & ketones	++
Aliphatic hydrocarbons	++
Aromatic hydrocarbons	++
Sugar alcohols & acids	++
Phosphoric acid	+

Note: ++++>1000ppm, +++>100ppm, ++>10 ppm, +>1ppm
(ppm = parts per million)
[See Note 25; permission Royal Soc.]

In 2011, Callahan and coworkers investigated the abundance of nucleobases of 12 different meteorites, including the Murchison, and detected adenine and guanine, along with their analogs. The distribution of nucleobases is similar to what was produced in the laboratory by reacting hydrogen cyanide, suggesting a mechanism for the origin of these extraterrestrial compounds.[16]

Comets are small, icy bodies that orbit around the sun with certain periodicities. The Comet "Wild 2" is believed to have formed at the high temperature region near the newly formed sun, and may have included particles from interstellar space long before our solar system existed. In 2006, NASA's spacecraft called *Stardust* was sent to snatch a sample of the comet and return it to Earth. Aside from inorganic materials, a host of organic molecules, believed to have been preserved for 4.5 billion years, was found. Mission scientists detected polycyclic aromatic hydrocarbons (PAH) and numerous nitrogen- and oxygen-rich compounds, including traces of glycine and beta-alanine (the latter not a protein component).[17] Some organic compounds found in comets are listed in Table 3.6.

Cosmic dust consists of tiny particles, ranging from a few molecules to aggregates of 0.01 mm in size, floating ubiquitously in space, frequently

Table 3.6. Some Organic Compounds in Comets

Name	Formula
Methanol	CH_3OH
Formamide	$HCONH_2$
Methane	CH_4
Ethylene	C_2H_4
Methylacetylene	CH_3C_2H
Formic acid	$HCOOH$
Acetonitrile	CH_3CN
Methyl formate	$HCOOCH_3$
Acetylene	C_2H_2
Ethane	C_2H_6
Hydrogen cyanide	HCN

[See Note 25; permission Royal Soc.]

Fig. 3.10. Infrared image of Trifid Nebula in the constellation Sagittarius. Nebular clouds like this are made of tiny interstellar particles called cosmic dust, and are believed to be nurseries for new stars. Cosmic dust contains simple organic molecules necessary for life. [US Natl. Aero. Space Ad.]

appearing like smoke or nebulous clouds (Fig. 3.10). Cosmic dust and clouds are the intermediates in the endless life-death cycles of the stellar world. Spectroscopic analysis of cosmic dust revealed the presence of many simple organic compounds containing carbon, oxygen, and nitrogen, in addition to hydrogen. Heavy elements are also present. Many of these are potential starting materials for the complex molecules of life (Table 3.7).

Next, can we simulate in the laboratory what went on in nature, starting from the molecules found in outer space or the early Earth? In 1953, Stanley Miller and Harold Urey produced a trace of amino acids upon concocting for two weeks a mixture of water, hydrogen, ammonia and methane, under constant electric spark discharge. At least 13 amino acids (mainly glycine and alanine) that can support life were identified in the mixture, amidst a host of unknown materials.[18] No nucleobases were found. Although the yield of the amino acids was minuscule, and the experimental conditions were subsequently considered not in conformity with those of early Earth (too much hydrogen), the experiment opened the door to the possibility of abiogenesis of life.

In 1961, Joan Oró synthesized adenine (a purine nucleobase) out of hydrogen cyanide (HCN), ammonia and water.[19] In a summary article

Table 3.7. Some Organic Molecules in Cosmic Dust

H_2	Hydrogen
O_2	Oxygen
CO	Carbon monoxide
CO_2	Carbon dioxide
H_2O	Water
OH	Hydroxyl radical
CH_4	Methane
CH_3OH	Methanol
H_2CO	Formaldehyde
NH_2CHO	Formamide
HCOOH	Formic acid
CH_3CHO	Acetaldehyde
CH_2OHCHO	Glycolaldehyde
CH_3COOH	Acetic acid
CH_3C_2H	Methylacetylene
H_2S	Hydrogen sulfide
OCS	Carbonyl sulfide
CS	Carbon sulfide
SO	Sulfur monoxide
H_2CS	Thioformamide
NH_3	Ammonia
CN	Cyanogen radical
HCN	Hydrogen cyanide
HC_3N	Cyanoacetylene
CH_3CN	Acetonitrile

As of 2012, there are over 200 simple molecules found in the interstellar and circumstellar clouds (in the form of dust and gas), as determined by spectroscopic analysis. About 20% of the dust consists of organic carbon compounds. [US Natl. Aero. Space Ad.]

published in 1967, Oró identified the following compounds as products of a simulated prebiotic synthesis: purines (adenine and guanine); pyrimidine (uracil); ribose and deoxyribose. (What is missing is cytosine which is unstable, decomposing quickly to uracil; ribose is also unstable unless complexed with borate.) The nucleobases were obtained by reacting hydrogen cyanide with cyanogen and cyanoacetylene, whereas the pentose monosaccharides were obtained by reacting formaldehyde. The same author also identified a number of amino acids as reaction products of hydrogen cyanide, ammonia and water: glycine, alanine, aspartic acid, serine, glutamic acid, threonine, leucine, isoleucine, arginine, with the first four predominating.[20]

3.7 Issue Number One: Which Came First?

This is a classic chicken-or-the-egg problem. There are three main macromolecular players that are considered of prime importance in a cell: DNA, RNA and protein. We also know that in a modern cell, the direction of information flow is from DNA to protein through RNA as an intermediary, but not the reverse. But this probably was not the case in the beginning of life. Though important as a depository of information, none of the functions of DNA (replication and information transfer) can happen without the aid of protein enzymes. The current consensus is that DNA probably was not the first macromolecule in the genesis of life. The choice then turns into RNA versus protein. Over the past few decades, attention has been focused on RNA as the choice candidate because of the discovery of "ribozymes," a type of RNA molecules that possesses enzyme (catalytic) activity.[21] This leads to the attractive possibility that RNA can both be the information carrier and the workhorse (enzymes) at the same time, before the advent of protein and DNA. A hypothesis of "RNA World" was proposed suggesting that there was an era in early life during which RNA "ruled the world." In this scenario, RNA becomes both the chicken and the egg.[22] In the following sections I shall evaluate the strengths and weaknesses of this theory.

3.8 Evidence in Favor of an RNA-first Scenario

These are the reasons supporting an ancient RNA World:

(1) RNA undergoes base pairing as does DNA, a prerequisite for template replication.
(2) Unlike DNA, which forms a rigid double-helix structure in a predictable manner, an RNA chain can fold upon itself by intra-chain

base-pairing, producing a 3-dimensional configuration unique for each RNA molecule. The ability to form a unique conformation is a prerequisite for enzyme catalysis.

(3) Many extant enzyme cofactors (coenzymes) are fragments of RNA and are suspected to be vestiges (molecular "fossils") of ancestral ribozymes. Coenzymes are small organic molecules that aid in the function of protein enzymes. Many coenzymes are RNA nucleotides or derivatives of these nucleotides. They include NAD (nicotinamide adenine dinucleotide), FAD (Flavin adenine dinucleotide), S-adenosylmethionine, coenzyme A, thiamine pyrophosphate, tetrahydrofolate, and pyridoxal phosphate. The implication is that RNA played an enzymatic role in early metabolism.[23]

(4) ATP (adenosine triphosphate), the universal energy carrier of a cell, is a derivative of ribonucleotide.

(5) Cyclic AMP and cyclic GMP, both derivatives of ribonucleotides, are important second messengers involved in intracellular signaling.

(6) Many ribonucleoside phosphates are used as intracellular activators for synthetic purposes: AMP, UMP, UDP, CMP, CDP, GMP, and GDP. (The abbreviations, respectively, are: adenosine monophosphate, uridine monophosphate, uridine diphosphate, cytidine monophosphate, cytidine diphosphate, guanosine monophosphate, and guanosine diphosphate.)

(7) In today's ribosome, the formation of peptide bonds of a protein is the work of a ribozyme.

(8) In the laboratory, random-sequenced RNA can be "crafted" to turn into ribozymes under man-made selective pressure. These man-made ribozymes can carry out specific catalytic functions (though less efficiently), including template-directed RNA replication, RNA aminoacylation, nucleotide synthesis, and peptide bond formation.[24]

(9) The clay montmorillonite has been found to be able to catalyze the formation of short RNA chains by polymerization of ribonucleotides.[25]

(10) Using cyclic AMP and cyclic GMP as starting materials, polymers (up to 25 nucleotides long) of adenylic and guanylic acids simulating

RNA chains have been formed under relatively mild temperature (40–90°C).[26]

3.9 Evidence against RNA: (I) Making an RNA Molecule Proves Daunting

Careful analysis reveals serious problems facing the RNA World hypothesis, foremost among which is the difficulty of making an RNA polymer without the help of protein enzymes:

(1) Although synthesizing ribose sugar in the laboratory is not hard, the newly formed ribose is unstable. Ribose can be made by reacting formaldehyde and glycolaldehyde, both found in interstellar space and probably on early Earth. It can also be formed by ultraviolet light irradiation of a mixture of water (ice), methanol and ammonia, all present in the formative stage of the solar system.[27] However, ribose decomposes very quickly unless it is stabilized by complexing with borate, a compound which is unlikely to be present in abundance in the prebiotic environment.[28]

(2) How were nucleobases synthesized spontaneously? Although purines are comparatively simple to synthesize chemically, the pyrimidine cytosine poses an insurmountable problem. Cytosine has never been detected in meteorites, and it is not produced in spark discharge experiments. The proposed prebiotic precursors (cyanoacetylene, cyanoacetaldehyde, cyanate, cyanogen and urea) were unlikely to be available in sufficient concentrations, and they were more likely to undergo side reactions with other organic compounds than form the desired product. And finally, cytosine, if formed, was too unstable (hydrolyzed to uracil) to last long enough over the alleged geological time for its incorporation into RNA.[29]

(3) Under plausible prebiotic conditions, joining nucleobases with ribose is fraught with problems. The reaction with purine nucleobases is extremely difficult and the reaction with pyrimidine nucleobases

does not work at all in the laboratory.[30] A number of experiments have been designed to bypass the obstacle. For example, Sutherland's group employed a highly tedious and complicated procedure to obtain cytosine and uracil nucleotides, proving in principle that activated ribonucleotides can be obtained without directly reacting a nucleobase and ribose.[31] Nevertheless, many of the critical steps require alternating exposures to mutually incompatible conditions such as heating/cooling and dehydration/rehydration; these highly precise and sequential steps were unlikely to occur in nature.

(4) The phosphodiester bond forming the backbone of the RNA molecule is unstable in water and easily hydrolyzed. This is the so-called "water paradox," a situation referring to the fact that biochemical reactions need to take place in an aqueous *milieu*, yet water is toxic to the biopolymers. In modern life, damage done by water to RNA is enzymatically repaired, but this mechanism was lacking in the primordial world where protein enzymes are absent. It was found that the solvent formamide (present in interstellar dust) stabilizes nucleic acids. However, it is doubtful that formamide could be present in such abundance on early Earth as to be able to support an RNA world.[32]

To sum up, despite the elegance of the dual-functionality of ribozymes, the major drawback of placing an RNA world ahead of others lies in the *difficulty of explaining the origin of RNA* itself.[33] To overcome the impasse, some investigators proposed the presence of "pre-RNA" replicative molecules that gave rise to the ribozymes, in effect negating the primacy of RNA in the origin of life. Following are some of the pre-RNA hypotheses: (a) Cairn-Smith proposed that the first replicating system was inorganic. Clays form naturally from silicates in solution, and clay crystals preserve their external formal arrangement as they grow. The irregularities of the distribution of cations (positively charged ions) could serve as the repository of information. Replication would be possible if any arrangement of the cations in a layer of clay dictates the formation of a new layer with identical charge distribution. Clay crystals attract certain molecules to their surfaces and turn them into complex organic molecules. Eventually the organic molecules (such as nucleic acids) acquire the ability to replicate. However, it is not clear how the last step could be accomplished.[34] (b) Eschenmoser synthesized an RNA-like molecule called pyranosyl RNA (pRNA) in which the ribose is replaced by a six-member ring analogue; the artificial molecule can potentially undergo standard Watson-Crick pairing mechanism of replication.[35] Others proposed replacing ribose with threose,[36] or ribose with glycerol.[37] The conversion from these nucleic acid analogues to RNA has not been demonstrated. (c) Using computer modeling, Nielson designed a polymer having a protein-like backbone and the usual nucleobase side chains, called peptide nucleic acid (PNA). Each strand of PNA is potentially capable of serving as template for the construction of its complement strand.[38] However, not only is the spontaneous synthesis of PNA questionable, the transition from PNA to RNA has not been explored.[39] (d) In an even more outrageous scenario, Francis Crick hinted that ready-made RNA (or ribozymes) could have been delivered to Earth from outer space, as part of his "directed panspermia" hypothesis.[40]

3.10 Evidence against RNA: (II) How did Functioning Ribozymes Emerge from Random-sequenced RNA?

Without artificial selection guided by a highly intelligent experimenter, it would be next to impossible to randomly pick an RNA chain having the correct sequence that fits a particular enzyme function. Take, for instance, an RNA molecule of 100 nucleotides long (a shorter one would compromise the fidelity of replication). Such a molecule would have 4^{100} (or about 10^{60}) possible random sequences. A pool of one copy for each sequence would weigh 10^{13} times the mass of the Earth. The odds of picking the right copy would be one in 4^{100}.[39] It is hard to imagine how this could have happened randomly in nature. Nevertheless, assuming that well-formed RNA enzymes were present pre-biotically, what would they do? First and foremost, they should also be able to replicate themselves. But since no self-replicating ribozyme has ever been found among present-day RNA, the attempt to simulate its existence in the ancient world can only be carried out in the laboratory.

In 1993, Bartel and Szostak generated a great deal of excitement by demonstrating that, from a mixture of RNA with random sequences, a species of RNA can be selected in test tube to perform template-directed polymerization.[41] Although such notable progress represents proof of principle, the artificially "evolved" RNA enzyme was obtained under conditions that could not have been provided by nature on the early Earth. Some of the problems are listed below: (1) The reactions were carried out in a well-protected, confined environment (test tubes or other reaction containers) with pure ingredients, whereas in the real world all reactions were exposed to a chaotic chemical mix, subjected to unlimited dilutions and multiple side reactions. (2) The reactions were carried out in a highly specific sequence with highly precise timing, requiring repeated purifications, isolations, and amplifications. (3) A "relay running" procedure was followed in which no starting reagent sees its way through the end product. By this I mean that once a desired product (however small the amount) is identified in each intermediate step, the experimenters purchased (from a chemical company) a large supply of that product in pure form and used it for the next reaction step. (4) Some enzyme reagents used were obtained from pre-existing life forms, such as transcriptase, reverse transcriptase, and DNA polymerase. (5) Some steps required sophisticated equipment available only in modern technology, such as the apparatus for polymerase chain reaction (PCR) and affinity columns utilizing biotin-streptavidin pairing. In short, the "selection" process was planned by highly intelligent experimenters and carried out in a modern laboratory setup, not by nature in a primeval environment. Yet, in spite of this degree of sophistication, the replicative function of the artificially generated ribozyme was slow and error-prone, and the copied segment was of limited length. Thus, the spontaneous generation of a self-replicating ribozyme has not been proved.

Setting aside the issue of replication, other artificial ribozymes have been produced to mimic a variety of unexpected enzyme functions. Again, there is no evidence that such ribozymes actually exist in nature, today or in ancient times. We should be cautious of the fact that the presence of a man-made compound or a man-made enzyme does not entail its presence in nature. For example, protein enzymes have been artificially produced that perform much more efficiently than their natural counterparts, although such artificial molecules do not occur in nature.

3.11 Evidence against RNA: (III) How did an RNA World Turn Over to a Protein World?

Perhaps the greatest conceptual hurdle encountered in formulating an RNA-dominated cell is how to visualize a smooth transition from an RNA-based to a protein-based metabolism without seriously disrupting the cellular order.

Theoretically, in a fully developed RNA World, there should be two pools of RNA: a gene-like RNA pool for information transfer and an enzymatic RNA (ribozyme) pool for metabolic function.[22] Furthermore, the 3-D conformation of the ribozyme, which is critical for its catalytic activity, is dictated by its nucleotide sequence, which in turn depends on sequence information contained in the genomic RNA. But once the enzymatic RNA was replaced by a protein, the information stored in the genomic RNA, which originally translated into a functioning ribozyme, no longer translated into a protein whose amino acid sequence formed a 3-D structure suitable for enzyme action. Thus, the transition from a well-formed RNA World to a Protein World as we see it today would amount to a major overhaul of the entire genetic and metabolic systems, possibly incurring deleterious if not devastating outcomes. This applies even to a minimally functioning "ribocyte" (a hypothetical RNA-dominated cell), which must contain at a minimum 100 genes of at least 40-nucleotides long, for the cell to be able to replicate and to start a primitive translation mechanism.[42] Because of the difficulty in transition, it is hard to imagine an RNA World as a precursor of a modern Protein World. Recently, scientists are starting to back down from a *pure* RNA World in favor of a Ribonucleo-Protein (RNP) World, in which RNA and peptides functioned in concert early on.[43]

3.12 The Pros and Cons of a Protein-first Scenario

The temptation in favor of a protein-first probiotic world is obvious. Proteins are versatile biopolymers that can perform a great number of tasks. Enzymes, in particular, are the universal workhorse of a cell. Once enzymes are in place, just about anything could happen. Furthermore, some amino acids occur naturally in meteorites that are as old as the birth of the Earth. However, although small peptides can occur spontaneously, big proteins can only be made in the ribosome through the action of *peptidyl transferase*, a ribozyme. Theoretically, the difficulty of making a functioning protein with the correct amino acid sequence far exceeds that of nucleic acids, as there are twenty different amino acids to choose from compared to only four nucleotides. Therefore, getting an active enzyme by random assemblage of amino acids is impossible.[44] Another major drawback with protein is that the molecule itself cannot serve as a template for replication, nor can it be used for any other kind of information transfer.

3.13 In Search of a Confined Environment

As noted above, metabolism takes place optimally in a privileged environment within an enclosure (the plasma membrane), which in turn needs metabolism and energy supply for its synthesis and maintenance. Here we hit another "chicken-or-the-egg" problem in addition to the one between proteins and nucleic acids. All laboratory simulations for the origin of life are conducted in test tubes, and it is all too easy to forget that nature did not provide such special containers when life was starting to germinate. Without a confinement, all the reactants would have been infinitely diluted and the chemistry of life could not have taken place, since the rate of a chemical reaction is concentration dependent (the law of mass action). A number of possibilities have been suggested, ranging from Darwin's "warm little pond" to lagoons, to volcanic hot springs, and to deep-sea hydrothermal vents, but none is considered satisfactory. Making a cell membrane requires lipids, which normally are made by enzymes. A ray of hope comes from the possibility that certain types of clay could catalyze the formation of fatty acids, and that clay could also convert fatty acids into cell-like vesicles.

Szostak and his coworkers showed that, in the test tube, simple fatty acids can form cell-like spheres that permit small molecules (such as amino acids and nucleotides) to get in, but prevent long RNA chains to diffuse out. Adding more fatty acid molecules to the mix permits the spheres to grow in size. Subsequent shearing of the larger spheres causes them to break apart into smaller ones, in a sense simulating cell division.[45]

3.14 The Power of Mineral Catalysis

Many metals are known catalysts for inorganic and organic reactions. In modern biochemistry, metal ions are essential components (serving as cofactors) in many enzyme systems. Examples include magnesium, manganese, copper, iron, nickel, zinc, cobalt and molybdenum. These and other metals are found in nature in the form of clays and were likely to be available in the early Earth. It is reasonable to assume that, before the advent of macromolecular catalysts (protein enzymes and ribozymes), metal ions could have ignited the first spark of life. Minerals in clays could serve as centers for carbon fixation and for anchoring growing carbon chains, promoting localized reactions in an otherwise chaotic chemical mix. Such immobilization would solve the dilution problem for the biopolymers, though it would not control other environmental factors such as pH, ionic strength, and the availability of small molecules.

Earlier, Cairn-Smith pointed out the importance of clay as a potential cradle of life.[34] Gunter Wachtershauser suggested that life started on the surface of a mineral of iron sulfide called pyrite, which has a positive charge, being able to attract pre-biotic negatively charged molecules, since most biomolecules are anionic.[46] He later proposed that this took place on the seafloor near the hydrothermal vents. On these sites the reducing power of the volcanic exhalations (H_2S and NH_3) led to carbon fixation on an inorganic substratum of catalytic iron and other transition metal centers. The pioneer metabolism could become autocatalytic and build up a layer of organic superstructure on top of the inorganic substructure on the seafloor.[47] A general review on the hypothesis of submarine hydrothermal vents as the nurseries of life has been published.[48]

It has been demonstrated in the laboratory that mineral surfaces promote the polymerization of nucleotides and amino acids. For instance, polyadenylates containing more than 50 monomers were formed on the clay *montmorillonite*, whereas polyglutamates of 55 monomers were formed on *illite*.[25,49] The results hint at the spontaneous formation of small RNA and proteins. Montmorillonite is a type of clay consisting of stacks of sheets, each made up of oxygen, aluminum, silicon, magnesium, iron and hydroxyl ion, arranged in regular arrays. It is formed by the weathering of volcanic ash and is likely to be present on the primitive Earth. Interestingly, montmorillonite has the ability to catalyze the conversion of fatty acid micelles into cell-like vesicles, a prototype of plasma membrane.[50] Thus, montmorillonite takes on an added significance in the origin of life.

In recent papers, it was demonstrated that, in the presence of mineral catalysts, a host of life-relevant molecules can originate from a single compound — formamide (H_2NCOH) — believed to be present on the primeval Earth. At temperatures of 90–160°C, and in the presence of a variety of minerals (silica, alumina, $CaCO_3$, TiO_2, zeolites, common clays, kaolin, montmorillonites, olivines and phosphate minerals), formamide condenses into a large number of nucleobases (including adenine, hypoxanthine, cytosine, uracil, thymine; excluding guanine which requires UV-irradiation). Each mineral catalyst promotes a specific panel of products. Of note is the finding that (1) acyclonucleosides are produced from formamide by TiO_2 catalysis, (2) nucleosides (adenosine and cytosine) can be phosphorylated by phosphate minerals, and (3) oligomerization of phosphorylated adenosine can spontaneously take place in formamide and water.[51]

3.15 A Case for Peptide/Nucleotide/Mineral Interplay

Perhaps it may not be appropriate to ask "Which came first?" regarding life's origin. After all, RNA and proteins need not be mutually exclusive. Why not visualize a scenario in which precursors of RNA and proteins developed interactively in early metabolism? As I argued above, an exclusive RNA World as a precursor for Protein World is untenable. In modern biochemistry, almost all catalytic reactions within a cell require three players: a protein (polypeptide), a fragment of RNA (a nucleotide cofactor), and a metal. Why not view this three-part cooperation as how life originated, instead of singling out the nucleotide cofactor as a "molecular fossil" of an extinct RNA World? The cooperation of peptides and oligo-nucleotides in early metabolism was suggested by Copley, Smith and Morowitz.[52] This scheme is also reminiscent of Calvin's idea of an early appearance of monomeric metabolites before their parallel polymerization into proteins and nucleic acids.[53] de Duve also stressed that bio-energetics should precede bio-informatics.[54] This proposal has been bolstered by the experimental finding of Sutherland and his associates, who showed that, from a simple "primordial soup" containing hydrogen cyanide and hydrogen sulfide (both ubiquitous in the universe), and under catalysis provided by ultraviolet light and copper, a list of precursors of the three main types of biomolecules — RNA, amino acids, and lipids — can be made.[55]

A plausible scenario may go like this. Let us start with amino acids, which have weak catalytic activity like some other organic compounds.[56] In the prebiotic world, these activities, in concert with metal ions, could be the first spark in the engine of life, promoting changes in other molecules, including synthesis of mono- and di-saccharides, and probably nucleosides, nucleotides, and di- and tri-nucleotides. Some of the amino acids might spontaneously link together through the amide bond,[57] making available di- and tri-peptides, further improving their catalytic function.

Over time, oligomers of amino acids and of nucleotides could appear. These snippets of peptides and RNAs did not have to be more than ten units long, and they might not contain all the subunits observed in today's biochemistry. For example, they might have only four simple types of amino acids (e.g., glycine, alanine, serine, aspartic acid), and two or three kinds of nucleobases (e.g., adenine, guanine and uracil). At this early stage, base pairing of RNA was probably unimportant, but it could be present. These oligomers of amino acids and nucleotides could undergo mutual or reciprocal catalysis (with broad substrate specificity) or, more significantly, form complexes (metal ions included) with improved catalytic function — a prototype of the modern day "protein/organic-cofactor/metal-cofactor" tripartite metabolic enzyme system. As metabolism became increasingly elaborate, more varieties of building blocks were available in the repertoire for incorporation into the polymers. At the same time proteins and RNAs could increase in length and begin to fold into 3-dimensional structures, greatly enhancing their respective enzyme functions. The proto-proteins and proto-RNA were initially confined to metabolic activities, but somewhere along the line complementary strands of RNA could appear, starting rudimentary molecular replication. Initially, the sequence of amino acids and nucleotides were largely random, and any RNA replication, if present, would be imprecise. How the useful sequences of the biopolymers were selected is a mystery, but I believe it could not have taken place without involving the organizational power of the proto-cell, which by this time should have been in the formative stage.

A plasma membrane was essential for the proto-cell. It was speculated that the cell membrane started as lipid vesicles or liposomes, to be replaced later by a phospholipid bilayer (as we have it today), and much later to be embellished with ion pumps and receptors, both of protein nature. The first membrane would be imperfect, permitting ions, amino acids and nucleotides/nucleosides to pass through but trapping biopolymers inside. The permeability of the crude membrane to ions suggests that the early cells might have their internal environment similar to the outside, implying that life started in terrestrial ponds adjacent to volcanic geothermal outlets, where the salt composition is closer to modern intracellular fluid than is seawater.[58]

Getting back to the RNA-protein relationship, a key molecule is *peptidyl transferase* (a ribozyme), which sits at the interface of RNA and protein worlds by making peptide bond connections between amino acids. The primitive transferase could well be much simpler than the modern counterpart, and might even function in the absence of a ribosome. Though grossly inefficient, this rickety machine should be able to "crank out" a few strands of long peptides (or short proteins), thus starting the protein "ball" rolling. Therefore, as soon as the *transferase* appeared, proteins could be made with relative ease, allowing them to gradually take over the metabolic processes. At the start, the proteins could be, for illustrative purposes, from 10 to 20 amino acids long, gradually increasing in length over millions of years. With improved efficiency of *peptidyl transferase* and the help of a proto-ribosome, the protein product could reach a length sufficient to fold on itself, forming 3-dimensional structures that enhanced their catalytic function. At the same time, proteins can help produce and stabilize long chains of RNA.

3.16 The Origins of the Genetic Code and the Translation Machinery

Despite much speculation and intensive experimentation, the origins of the genetic code and its assignment, along with the translation mechanism, are shrouded in mystery (Table 3.3; Figs. 3.3 and 3.4). Based on our knowledge that the code is made of three letters, each taken from an alphabet of four nucleobases, there are $4 \times 4 \times 4 = 64$ possible codons.

But since there are 20 different kinds of amino acids, if we were to randomly associate all the amino acids with all the available codons (giving allowance for one stop codon; a stop codon is like a "punctuation mark" at the end of a protein "sentence"), we would end up with approximately 1×10^{84} different versions of the coding system (derived from 21^{64}). That means for two independently arisen genetic code systems to be identical, the chance is one in 10^{84}, a virtual impossibility.[59] The fact that all extant organisms on Earth use the same standard genetic code indicates that the choice cannot be random. This, along with the unity of energy utilization and biosynthetic pathways, provides a strong argument for a common ancestry for all lives on Earth. Alternatively, if life sprang up from more than one site, there must be certain deterministic factors involved in the process.

The following information suggests the presence of deterministic factors in the assignment of the genetic code: (1) Codons for the same amino acids, or for amino acids with similar physical properties, are grouped together so that a mutation by changing one nucleotide is less likely to alter the property of the protein. For example, multiple codons for the same amino acid all have similar first two nucleotides; codons for aspartic acid and glutamic acid differ by only one nucleotide (the last); amino acids with a hydrophobic side chain all have the nucleotide U in the middle of the codon. (2) Amino acids that share the same synthetic pathway tend to have the same first nucleotide in their codons.[60] (3) The four amino acids that are the most abundant in meteorites and in Miller's prebiotic synthetic experiments — glycine, alanine, valine and aspartic acid — are coded by the codons GCC, GGC, GUC and GAC, respectively, the only variation being in the second letter, implicating that the primitive codon started with the triplet GXC, with X being one of the four nucleotides.[61] (4) Experiments on *in vitro* RNA-amino acid binding found that, for the majority of amino acids, the RNA binding site usually contains either a codon or an anticodon corresponding to that amino acid.[62] Based on this finding, Yarus proposed a stereochemical hypothesis for the origin of the genetic code. He posited that the current triplet genetic code is a vestige of the amino acid-tRNA affinity in ancient times. The question arises as to why stereo chemical interactions no longer apply to the modern tRNA. Yarus explained this with another hypothesis, called "escaped triplet hypothesis," which says that in later days the original binding triplets were relocated to become codons on mRNAs, and anticodons on tRNAs.[63] How the relocation took place has not been explained.

It would be impossible to consider the origin of the genetic code without also tackling the overall translation mechanism — the conversion of the code from the RNA language to the protein language. In modern cell biology, there are four major players in this information transfer: (1) the activated free amino acids (to be incorporated into proteins), (2) the transfer RNA (tRNA) that contains the anticodon, (3) the messenger RNA (mRNA) with its three-letter codons, (4) the protein enzyme aminoacyl-tRNA synthetase (AARS). (Peptidyl transferase is part of the ribosome, but it is not involved in message interpretation.) tRNAs are adapters that match the cognate amino acids to the three-letter genetic codes on the mRNA. This is accomplished in two steps. The first is the attachment of amino acids to the correct tRNAs, aided by aminoacyl-tRNA synthetase. The amino acid is hooked to one end (the 3' end) of the tRNA molecule with a covalent bond, while near the middle of the tRNA is situated the anticodon (a triplet-nucleotide), to be bound to the complementary codon on the mRNA in the next step. AARS is unique in that it serves as an interpreter that "speaks two languages" — RNA and protein. The enzyme is able to do this by possessing dual recognition sites (see Fig. 3.4A), one for the amino acid and the other for the anticodon on the tRNA. AARS is a family of 20 different proteins with considerable structural diversity, each recognizing a particular amino acid; but because of degeneracy of the genetic code, each amino acid may correspond to more than one codon, and thus requires more than one cognate tRNA. In fact, there are 61 different tRNAs, corresponding to the number of available genetic codons (64 minus 3 stop codons in humans), whereas there are only 20 AARS corresponding to

20 amino acids. The discrepancy between the number of tRNAs and the number of AARSs has to be solved. Put it another way, how can 20 AARS attach 20 different amino acids to 61 different kinds of tRNAs and come up with a correct match (obviously not a one-to-one match)? For example, the amino acid serine is coded by six different codons, some of which bear no similarity at all. The single AARS for serine is able to bring the amino acid to all six different tRNAs assigned to serine, despite variations in the anticodon in these tRNAs. The explanation is that only some of AARSs recognize the anticodons on the tRNAs; the others recognize structural features in the tRNA molecules other than the anticodons, so that their matching task is unaffected by changes in the anticodons. Therefore, for AARS, the recognition of the general molecular structure of a tRNA takes precedence over its anticodon. This fact is strong evidence that protein-RNA interaction came before genetic coding in the origin of life. (The remarkable matching ability of an AARS can be compared to that of a clever matchmaker who recognizes a boy not only by his legal name but also by all his five nicknames, so that no identity error is committed when attempting to pair him with a certain girl.)

The last step in translation is the binding of the amino acid-charged tRNA to mRNA through codon/anticodon matching. Once multiple tRNAs are lined up along the messenger RNA on the ribosome, the simple task of joining the amino acids together is performed by the peptidyl transferase in the ribosome.

3.17 Speculations on the Primitive Translation System

The modern translation machine is a highly sophisticated system. It would be reasonable to assume that the primitive counterpart would be much simpler and might even be drastically different. For example, we can imagine that the 3-letter genetic code might have only two nucleobases to choose from (instead of the current four), generating only eight possible "words," permitting eight amino acids to be coded for. The peptide length could be short and variable, and the coding could be imprecise and highly "wobble". Thus, amino acids of related properties might substitute for each other, and the proteins might function enzymatically in a statistical rather than discrete manner.

The primitive tRNA might serve as its own amino acid activating enzyme, before the advent of aminoacyl-tRNA synthetase (AARS).[64] Along this line of thinking is the hypothesis of "proto-anticodon RNA," which posits that amino acids stereochemically bind to their respective anticodons that are nested in the proto-tRNA; the proto-tRNA presumably serves as an enzyme catalyzing the aminoacylation of its own 3' end, forming aminoacylated proto-tRNA. This hypothesis attempts to explain (1) the origin of the genetic code, and (2) why only the "L" form of amino acids is incorporated into proteins. (The latter is due to the orientation of the amino acid side chains with respect to the space created by the anticodon and its adjacent nucleotide structure, allowing only the L-amino acids but not their "D" counterparts to fit in.) This hypothesis places anticodons ahead of codons, both preceding the translation mechanism.[65]

3.18 A Recapitulation: What does a Genetic Code Mean?

A code, like any symbol, has no meaning unless it signifies something. For those who subscribe to the RNA World hypothesis, it naturally follows that the information-laden RNAs (equivalent to the present day mRNA) existed before the codons were translated to the amino-acid sequence of proteins. However, as I argued above, a nucleotide sequence meaningful to an RNA World does not make metabolic sense when translated into a protein. It is more reasonable to believe that the genetic code appeared as a way to preserve the information derived from primitive proteins that were metabolically functioning. The primitive mRNA thus formed could double as a gene by undergoing replication by complementarily, either by autocatalysis, or by the action of ribozyme or protein enzyme.

The final step in the origin of life would be the deposition of genetic information in DNA molecules. DNA is superior to RNA for this purpose because of its stability, but this could not have happened before the reduction of ribose to deoxyribose, and the methylation of uracil to thymine. Therefore, the history of information transfer during the formative stage of life would be protein → mRNA → DNA, the reverse of the Central Dogma.

3.19 Issue Number Two: Spontaneous Organization

In addition to the question of which molecules came first, there is another issue in the origin of life, and that is spontaneous organization. That in the living system matter interacts to generate complexity is a given, but how this happens is puzzling. Complexity is an emergent phenomenon that cannot be explained by simple physical laws, including thermodynamics.[66,67]

Using computer modeling, Stuart Kauffman showed that a self-sustaining, autocatalytic process can be spontaneously generated from a group of interacting chemicals, provided that they are of sufficient diversity, and that the system is constantly supported with energy input. As Kauffman stated: "Life is a natural property of complex chemical systems. When a number of different molecules in a chemical soup passes a certain threshold, a self-sustaining network of reaction — an autocatalytic metabolism — will suddenly appear."[68] However, its application to real life is problematic, as we do not know which empirical data to select for the input. Science writer M. Mitchell Waldrop has the following description: "No matter how many calculations and computer simulations he (Kauffman) carried out on the origin of life, they were still just calculations and computer simulations. To make a really compelling case, he would have had to take the experiments of Miller and Urey one step farther, by demonstrating that their primordial soup could actually give rise to an autocatalytic set in the laboratory. But Kauffman had no idea how to do that. Even if he had had the patience and knowledge to do laboratory chemistry, he would have had to look at millions of possible compounds in all conceivable combinations under a wide range of temperatures and pressures. He could have spent a lifetime on the problem and gotten nowhere. No one else seemed to have any good ideas either."[69] Kauffman himself also conceded, "Central to this is a missing theory of organization…My difficulty in constructing such a theory is, at least, that I cannot see how to begin…Yet the biosphere, for 4.8 billion years, has been doing just this."[67] A verifiable law for naturally occurring complexity is yet to be found.

3.20 An Aside: A Model for Molecular Self-assembly using Synthetic "Organic Bricks"

In an experiment unrelated to genetics, a model for self-assembly has been demonstrated using single-stranded DNA as construction "bricks". DNA strands of 32 nucleotides long, containing four regions that can hybridize (by Watson-Crick base pairing) to four neighboring DNA strands, were synthesized and used like LEGO bricks. Arbitrarily prescribed 3-D structures were formed by selecting different subsets of strands from the same pool of DNA sequences. The generality and diversity of the method was demonstrated by creating 102 different structures out of a subset of $10 \times 10 \times 10$ voxel cuboid. Each voxel corresponds to 8-base pair interaction between bricks or the equivalent of a one-stud LEGO brick.[70] This interesting model shows that complexity can come from simple principles. Nevertheless, the model falls short of being spontaneous, as it was designed by human intelligence.

3.21 What have we Learned about the Origin of Life?

Over the last fifty years, substantial progress has been made on formulating hypotheses on the beginning of life on Earth. However, we are far from putting the pieces together to arrive at a coherent picture. Deep gaps still exist, preventing us from making sensible connections among isolated clues. Obviously it would be impossible to re-enact the entire process in the laboratory. Even if we are lucky enough to witness the genesis of life in some remote planets, our life span is far too short to chronicle the entire drama. The laboratory is a place to simulate, but the setup is all too artificial, to the extent that very little is natural and nothing can be considered spontaneous. Simulations can also be made with computer modeling, but models are not real, and the initial conditions and the algorithms can only be assumed. As of now, our knowledge of life's origin remains a fuzzy guesswork whose true nature may not be known for sometime. Some experts, at some point in their careers, expressed frustration:

> "An honest man, armed with all the knowledge available to us now, could only state that in some sense, the origin of life appears at the moment to be almost a miracle, so many are the conditions which would have had to have been satisfied to get it going." — Francis Crick[71]
>
> "Anyone who tells you that he or she knows how life started on the sere Earth 3.45 billion years ago is a fool or a knave. — Stuart Kauffman[72]
>
> "And so, at first glance, one might have to conclude that life could never, in fact, have originated by chemical means." — Leslie Orgel[73]

Unable to explain the origin of life, Darwin compared it to the mystery of gravity. He stated in 1861 in the 3rd edition of *The Origin of Species*: "It is no valid objection that science as yet throws no light on the far higher problem of the essence or origin of life. Who can explain what is the essence of the attraction of gravity?"[74] Certainly, we know a lot more than Darwin did 150 years ago, but the beginning of life on Earth, or elsewhere in the universe, continues to puzzle us today.

There is also the issue of chance versus necessity. Did life happen against all odds, an extremely unlikely accident that occurred perhaps only once in the universe's lifetime, as Monod suggested?[75] Or is it a cosmic imperative, something bound to be born no matter what, as de Duve[54] and Eigen[76] insisted? To this controversy, I can only say that cosmic imperative alone would not be sufficient; it rather sets the permissive and restrictive conditions for chance to strike. Life resulted from a combination of both, but how much each factor contributed, I cannot tell. The analogy is rather like a forest fire. It takes time to build up the right physical condition in favor of combustion, but once a spark starts the fire spreads rapidly.

Life is built on chemistry but it ends up being more than chemistry. When molecules group together under certain propitious conditions, they interact, interconnect and form complex, dynamic, hierarchical networks. It is then, and only then, that an assemblage of matter expresses a unified goal, or what appears to be a goal, to continue its own existence, including making more of itself. It is as if life, once born on our planet, announced itself with a determination to go on forever.

3.22 The Birth of Life is the Birth of *Self*

The Earth was formed 4.5 billion years ago, and the evidence of life dates back to about 3.5 billion years. The Earth in the first half billion years is considered to be too tumultuous to permit any living things to occur, so it is reasonable to assume that the process leading to the beginning of life started around 4 billion years ago. The question is, was the formation of life a gradual and protracted process, taking place over several hundred million years, or a relatively rapid one, occurring over a few thousand years? I believe the process that culminated in the appearance of life is gradual, arduous, chaotic, and error-prone, but the birth of the first life could have taken place within a watershed period, a relatively short geological time frame. The slow transition period would allow time for carbon based molecular chains to form, extend, and accumulate; some metal-based catalysis to take place; some short, random sequenced polypeptides and oligonucleotides to form (not using the whole gamut of modern-day amino acids and nucleobases); and some simple lipid vesicles to appear, coalesce, and split. We can call these potpourris of carbonaceous materials "pre-life", "pre-cell", "proto-life", or "proto-cell", but *they were not true life*. The first real life could be extremely simple, much simpler than any we can see today, but nonetheless it should be complete and capable of independent existence, which should meet the following minimal requirements: (1) it should be enclosed by a semi-permeable membrane; (2) it should show a net energy gain (taking in more than dissipating); (3) it should be able to defy the second law of thermodynamics — building up and maintaining an internal order amid a chaotic external environment; (4) it should be able to maintain relative stability long enough for reproduction to take place; (5) it should be able to reproduce with a fair degree of faithfulness.[77]

In more technical terms, a complete life should be all of below: self-assembled, self-organized, self-programmed, self-contained, self maintained, self-optimized, self-correcting, self-propelled and, lastly, be able to self-reproduce. If you are slightly annoyed by the litany of the prefix "self," it is because *self* and life are inseparable. The birth of life is also the birth of *self*.

(Please note that at this stage *self* and life are almost equivalent, but toward the end of the book when I deal with higher animals, especially humans, the concept of *self* extends to include the inner aspects of our being.)

3.23 Did Evolution Start Life?

Whereas Darwinian evolution is irrefutable when applied to the biological world, whether evolution was a driving force leading to the emergence of life is a different issue. Darwin himself insisted that his theory applies only to the succession of life, but not to its beginning, to which he professed ignorance.[78] Important knowledge has been accumulated since Darwin's time, causing some people to extrapolate evolution to the pre-biotic era.[79,80] Whether or not such extension of the Darwinian worldview is justifiable is the subject of this section.

Biological evolution is the change of characteristics of a population over successive generations. Charles Darwin's seminal contribution is to propose that such changes are the outcome of two factors: in inherent variation of organisms from generation to generation, and the external environment that allows new progenies that fit the environment to survive and propagate, giving rise to individual diversity, adaptation, and eventually the production of new species. This mechanism, called natural selection, is the theme of Darwinian evolution. "Selection" is possible because organisms produce more offspring than natural resources can support, and the survival of the fittest is a matter of competition.

Darwinian evolution received a boost in the early 20th century, when genetics was accepted and integrated into the theory. It was then realized that gene mutation is a major cause of variation across generations. The mechanism of mutation was explained in the mid-20th century by the recognition of DNA as the genetic material and the advent of the double-helical Watson-Crick model of DNA, which provides a molecular mechanism for replication and inheritance. The natural tendency to make occasional errors in copying a DNA chain is the reason why mutation spontaneously occurs. With this insight Darwinian evolution can be reduced from the biological to the molecular level — the so-called "molecular evolution". For example, a mutation that changes one nucleobase in the DNA can alter the protein sequence, making it more, or less, active as an enzyme, thus affecting the chance of survival for the organism.

However, I like to point out that molecular evolution makes sense only in the context of cellular life, for ultimately it is the survival of the cell that counts. Contrary to this idea, some people equate changes in the DNA molecule (or any other replicating molecule), under any circumstances, as molecular evolution in the Darwinian sense. They insist that, when extrapolated to the pre-biotic era, the speed of replication and abundance of a hypothetical DNA molecule, even in a cell-free situation, qualifies for evolution.

For the sake of argument, it will be helpful for me to recount an experiment performed by Spiegelman and his colleagues in the 1960s, considered by some to be a prototype of pre-cellular "molecular evolution".[81] Spiegelman and his colleagues studied the *in vitro* replication of an RNA virus called Q-beta. (Normally only DNA is capable of replication, but in Q-beta, RNA is the genetic material and it functions like DNA.) In the absence of cellular components, multiple copies of Q-beta RNA can be made in a test tube in the presence of a protein enzyme called RNA-replicase. Their experiment consisted of incubating the RNA with the enzyme (along with the required raw materials for RNA synthesis) and letting the RNA multiply in number. After a fixed period (say, 20 minutes), a minute amount of the mixture (containing the replicated RNA product) was transferred to seed another test tube, where a fresh preparation of the enzyme and the precursors were again added. Upon further incubation, a tiny amount of the newly replicated RNA was in turn transferred to a third tube for reaction with a new preparation of the enzyme and precursors. The transfer of tubes went on and on, and the experimenters shortened the time of incubation in each step. After many transfers, the final product was a mutant RNA that replicated many times faster than the starting RNA. What happened was that, when a replicator (RNA or DNA) divides, a copying error (mutation) normally occurs, at the rate of one per 100,000 nucleotides. Since the mutations are random, in this experiment some mutants would promote the replication rate, while others would not. By stepwise shortening the incubation time, the experimenters selected those RNA variants that multiplied faster than others, and this favored trait was accumulated over succeeding transfers. In the end, a fast replicating RNA came out the winner of the race. Spiegelman himself dubbed this process "Darwinian evolution."[82]

Let me analyze how well this experiment agrees with the Darwinian concept of evolution. On the positive side, the experiment shows the ability of genes (here represented by RNA) to change (mutation), and that some of these changes can be preserved by external selective pressure (here imposed by the experimenters). What it did not demonstrate is that the abundance of a replicating molecule leads to the survival of a cell (since there is no cell involved), let alone the making of one. A replicating molecule, taken alone, does not translate into a functioning protein useful for metabolic purposes. Piling up tons of DNA (or RNA) is not the same as making a live cell. At best, Spiegelman's experiment is

"Darwinian evolution" only in a very limited sense, not general enough to explain the emergence of life on Earth. In my opinion, applying the idea of Darwinian evolution to the pre-biotic era is an over-extrapolation of the principle, akin to putting the carriage before the horse.

Before life started, evolution had no substrate to act on. After cellular life appeared, natural selection acted relentlessly on life's yearning for continued existence, an expression of *self*. Evolution did not start life, but life, and *self*, "kicked off" evolution. Hence, to quote Darwin, "from so simple a beginning endless forms most beautiful and most wonderful have been, and are being, evolved."[83]

3.24 *Self* as the Driving Force of Evolution

Ever since the proposal of Darwinian evolution, the substrate of natural selection has been a moving target. First, it was the organism that was selected. Next, it became clear that the gene, which is responsible for variations in individual characteristics, was the unit of selection. By the mid-twentieth century, DNA was established as the genetic material. DNA contains information (through its coded sequence) that can pass on to future generations through replication, while the same information can specify the type of protein to be made. DNA thus became the center of life. It was on this basis that the "selfish gene" theory of Richard Dawkins was proposed in 1976.[84]

Dawkins distinguished two complementary entities in a living thing: the *replicator* and the *vehicle*. The replicator is the information-carrying molecule whose function is to perpetuate the information *ad infinitum* (barring occasional mutations); the vehicle is the rest of the organism whose role is to make possible the continued replication of the replicator. In other words, the replicator is the "boss", whereas the vehicle is the "servant." Only one type of molecule in the body qualifies as a replicator, and that is DNA; all others, including protein, are vehicles. In Dawkins' own words, "a vehicle is any unit, discrete enough to seem worth naming, which houses a collection of replicators and which works as a unit for the preservation and propagation of those replicators."[85] Evolution can then be reduced to a matter of competition among genes (or segments of DNA) for abundance.

While this elegant idea concurred with the then accepted knowledge of molecular biology and gained popular acceptance for several decades, a number of limitations gradually crept in and prompted re-examination of the concept. First, proteins and genes (if defined as segments of a DNA molecule) do not necessarily have a one-to-one correspondence. Only 2% of human DNA codes for proteins (qualifying as true genes); the rest serves regulatory functions. By a process called alternative splicing, segments from distant parts of a DNA molecule (even from different chromosomes) can be bought together to form a "paste-up" protein, while a given segment of DNA can find its way into different proteins with diverse functions. Thus, the information transfer from DNA to protein is scrambled, and DNA sequences turn out not to be exact templates for protein molecules. The adage "one gene, one enzyme," or "one gene, one protein", once thought to be infallible, no longer holds true. What then is DNA good for? Sure enough, it is a handbook, an encyclopedia, a reference source, a storehouse for information, and a blueprint suitable for copying, but it is not capable of replication or information transfer without the help of many proteins, let alone making a live cell.

The concept of *epigenetics*, which posits that inheritance is not limited to the information coded in the DNA sequence, is gaining acceptance. The mechanisms include methylation of DNA, modification of histone (the protein that binds DNA in the nucleus), and multiple versions of RNA interference and regulations, along with external factors such as culture and civilization that can pass on from generation to generation.[86] Inheritance has now expanded from the "one-dimensional" DNA to a "multi-dimensional" biological process.[87]

It may be worth mentioning that in certain species of yeasts, non-Mendelian inheritance from one cell to the daughter cells can be carried out through "prion" proteins, a type of protein that exists in two functionally distinct conformations, one of which, called the prion form, can convert other proteins of the same type to the prion state by forcing it to fold in a certain manner. By direct transmission of the prion protein to the daughter cells through the cytoplasm, the inheritance bypasses DNA mediation.[88]

One additional limitation of the "gene's-eye view" of Dawkins is the difficulty of explaining group selection. That is, why will animals, people included, of the same group help each other even if they are not close relations? For this reason, Dawkins vehemently rejected altruism in a group situation except among immediate relatives (kin selection) and in the case of reciprocal altruism, in which mutual benefit is apparent. By contrast, my "*self*'s-eye view" of life is consistent with the concept of *multilevel* selection, as I consider groups at any level to be an *expanded self*, all capable of competing other groups of the same nature and the same rank for perpetuation (to be elaborated in *Chapter 12*).

In short, the unit of natural selection, which once shifted from the organism to DNA or gene, is now switched back to the organism (with some modifications). The prevailing view is that an organism is a *dynamic system* that interconnects different parts, and DNA is just one of these parts. *Self* as defined in this book fits nicely in this new framework. At the bottom, it is *self*'s unwavering craving for *survival* that makes natural selection and evolution possible. *Self* is therefore the invisible axis around which life spiraled forward over billions of years of evolution, and the process continues without end.

3.25 On Purpose and Design

Nature does not have a purpose. Instead, science discovered laws that natural objects seem to "obey," and in obeying these laws natural events do show an endpoint (or a series of endpoints). A rock falling to the ground will roll to the lowest point and stop according to the law of gravity. Likewise, movements executed by organisms also have endpoints, but living things will overcome barriers or take alternative routes to reach them, a process we commonly call "behavior." For example, all organisms need to eat, but they can choose different ways to satisfy their hunger. Here, instead of an endpoint we see a *convergent* point for a set of alternative courses, or what appears to be a "goal." How this simple "goal-seeking" tendency of living matter, which initially was devoid of conscious accompaniment, would one day, following billions of years of evolution, turn into a full-fledged conscious purpose at the human level is indeed a great wonder of nature.

What about intelligent design? When Einstein spoke of the physical universe as understandable, he implied a plan, a design. When Walter Cannon named his book *The Wisdom of the Body*, he admitted to an intelligible scheme behind the workings of the body.[89] That life has an underlying common mechanism is supported by the uniformity of biochemistry, despite life's outward diversity. In the view of Thomas Kuhn, natural science works on the assumption of a basic framework — a paradigm.[90] A paradigm is useful as long as it is internally consistent. If we assume a designer who created the universe and left it alone once completed (*deist* view), it still makes sense for scientists to find out how the universe works. But if we assume a whimsical (though just and benevolent) supreme ruler who constantly intervenes in the workings of the world (*theist* view), and responds to people's diverse and often conflicting wishes, scientific exploration would be futile, since the paradigm loses its internal coherence and intelligibility. To call this "intelligent" design would then be self-contradictory.

An alternative to the deist view that is compatible with scientific outlook is to take the universe as a natural occurrence and not a handiwork of an intelligent, conscious being. Under this category would belong atheism and agnosticism; the difference between the two could easily be a matter of taste. Bertrand Russell said that if he were asked by a philosopher, he would call himself an agnostic, since there is no way to prove or disprove the presence of God. On the other hand, the absence of a disproof does not constitute a proof. So, if he were asked by a layman, Russell said he would just as well call himself an atheist.

The analogy is like the Olympian gods (Zeus and his clan), of which we have neither positive nor negative proof.[91]

3.26 Is Biology Reducible to Physics?

The topic of biological reductionism is plagued with confusion because different people use the term in different ways. In the most simplistic form, there is reductionism versus anti-reductionism, in which the former maintains that physical (and chemical) principles can *fully explain* all biological facts, whereas the latter attributes all workings of life to an intangible vital force, which in the extreme case would be a soul, something that gives life to matter but can exist independent of matter. Progress in biochemistry, molecular biology, biophysics, and physiology has proved the extreme vitalistic view wrong. On the other hand, gaps exist in the problem of self-organization at different levels of the living system, along with an incomplete understanding of the origin of life and the nature of mental phenomenon, leaving the full explanation of life an unaccomplished task. Thus, perhaps the best answer is not a blanket statement of "yes" or "no," but to assess how much has been reduced, how much has not yet been reduced, and perhaps what cannot be reduced. Please note that what cannot be reduced does not necessarily entail a spiritual vital force; it could simply be hitherto undiscovered natural laws applicable only to living things.[92]

There is elegance in simplicity and unity of knowledge, and for the sake of unity modern science strives to seek a connection between living things and inanimate objects. Life is the most complex entity in the universe. We can reduce this complex entity by taking it apart down to its tissue, cellular and molecular components and observe how each part works. From this downward approach, science can triumphantly claim that for every biological phenomenon there is a corresponding physical mechanism. However, the reverse process, that of putting the parts together to make a dynamic whole, is much harder. Until this upward process is understood, we have to consider life as an emergent, one that cannot be foreseen by principles observed on a lower level.[93,94]

The principle of biological organization (presumably a natural law), though innate to matter, does not manifest itself until the molecules are assembled in a certain manner under the most propitious conditions. Facing the enormous complexities of life, Martin Rees, the astrophysicist, cosmologist and President of the Royal Society of Britain, transpired a sense of humility: "Living things embody intricate structures that render them far more mysterious than atoms or stars. Will scientists ever fathom all of nature's complexities? Perhaps they will. However, we should be open to the possibility that we might encounter limits because our brains just don't have enough conceptual grasp. There seems no reason why human mental capacity should be matched to all deep aspects of reality."[95] The admonition from Rees is that there might be voids in reality that human knowledge can never fill. As to whether physics can explain all the phenomena of life, I like to borrow the chess playing analogy offered by Rees (alluded to Richard Feynman): To a person who has never seen chess being played, he can quickly grasp the rules by watching the game a few times. But knowing how the pieces move is just the first step on the long road from being a novice to a grand master, for the beauty of the game lies in the rich variety that the rules allow. Likewise, the richness and complexity of life might not all be inferable from the simple laws of physics as we know them today.[95,96]

Notes and References

1. Pauli W. Exclusion Principle and Quantum Mechanics, 1945 Nobel Lecture, Neuchatel, 1947.
2. The four quantum numbers are n, l, m and s, where n and l are related to the energy level of the orbital, m and s are related to the magnetic properties arising from the motion of the electrons; m from revolution around the nucleus, and s from self-rotation or spin of the electron.
3. This picture of bonding does not apply to metals, in which case all electrons are loosely connected with the atomic nuclei and behave in a collective manner.

4. Gravity plays a lesser role.
5. Of all the elements, the nucleus of iron contains the lowest amount of energy. Therefore, fusing lighter elements into iron releases energy, whereas fusing iron into higher elements requires net energy input.
6. Many of these signaling proteins have enzyme activities, consisting of adding a phosphate group (phosphorylation) to a molecule, or removing one (dephosphorylation). Enzymes that perform phosphorylation are called kinases; those that perform dephosphorylation are called phosphatases.
7. A steady state, in simple language, is constancy in a state of flux. As an example, a waterfall retains its shape and appearance despite rapid rushing of water.
8. Schrödinger E. (1944) *What is Life? The Physical Aspects of the Living Cell.* Cambridge Univ. Press, Cambridge, UK.
9. Quoted by Webster G, Goodwin BC. (1982) The origin of species: A structuralist approach. *J Soc Biol Struct* **5**: 15.
10. Although minor deviations from the standard code do exist in mitochondrial and chloroplast DNA, the uniformity is still overwhelming.
11. Schidlowski M. (1988) A 3,800-million-year isotope record of life from carbon in sedimentary rocks. *Nature* **333**: 313–318.
12. In a letter mailed to Hooker in 1871, Darwin stated: "It is often said that all the conditions for the first production of a living organism are now present, which could ever have been present. But if (and oh what a big if) we could conceive in some warm little pond with all sorts of ammonia and phosphoric salts — light, heat, electricity, etc., present, that a protein compound was chemically formed, ready to undergo still more complex changes, at the present day such matter would be instantly devoured, or absorbed, which would not have been the case before living creatures were formed." See: Darwin F. ed. (1887) *The Life and Letters of Charles Darwin.* Vol. 3, John Murray, London, pp. 168–169.
13. Botta O, Bada JL. (2002) Extraterrestrial organic compounds in meteorites. *Surveys in Geophysics.* **23**: 411–467.
14. Cooper G, Kimmich N, Belisle W, *et al.* (2001) Carbonaceous meteorites as a source of sugar-related organic compounds for the early Earth. *Nature* **414**: 879–883.
15. Martins Z, Botta O, Fogel ML, *et al.* (2008) Extraterrestrial nucleobases in the Murchison meteorite. *Earth and Planetary Sci Lett* **270**: 130–136.

16. Callahan MP, Smith KE, Cleaves HJ II, *et al.* (2011) Carbonaceous meteorites contain a wide range of extraterrestrial nucleobases. *Proc Natl Acad Sci USA* **108**: 13995–13998.
17. *Chemical and Engineering News* (2007), **85**: 41–42.
18. Miller SL. (1953) A production of amino acids under possible primitive earth conditions. *Science* **117**: 528–529.
19. Oró J, Kimball AP. (1961) Synthesis of purines under possible primitive earth conditions. I. Adenine from hydrogen cyanide. *Arch Bio Chem Biophys* **94**: 217–227.
20. Oró J, Kamat SS. (1961) Amino-acid synthesis from hydrogen cyanide under possible primitive earth conditions. *Nature* **190**: 442–443; Oró J. (1967) Stages and mechanisms of prebiological organic synthesis. In Fox SW. ed. *Origins of Prebiological Systems and of Their Molecular Matrices*. Academic Press, New York, pp. 137–171.
21. Unlike DNA chains which are rigid except to coil around and form a double helix, RNA chains are highly flexible so that the various regions of a single chain can come together by base-pairing to produce highly irregular, three-dimensional structures unique for each RNA molecule (reminiscent of protein folding). Examples of RNA enzymes include self-splicing RNA from the protozoan Tetrahymena, and RNAseP, a nucleoprotein whose RNA part is responsible for RNA-cutting activity (nuclease) during the formation of tRNA. See: Kruger K, Grabowski PJ, Zaug AJ, *et al.* (1982) Auto excision and auto cyclization of the ribosomal RNA intervening sequence of Tetrahymena. *Cell* **31**: 147–157; Guerrier-Takada C, Gardiner K, Marsh T, *et al.* (1983) The RNA moiety of ribonuclease P is the catalytic subunit of the enzyme. *Cell* **35**: 849–857.
22. Gilbert W. (1986) The RNA World. *Nature* **319**: 618.
23. White HB. (1976) Coenzymes as fossils of an earlier metabolic state. *J Mol Evolution* **7**: 101–104.
24. Bartel DP, Szostak JW. (1993) Isolation of new ribozymes from a large pool of random sequences. *Science* **261**: 1411–1418; Zhang B, Cech TR. (1997) Peptide bond formation by in vitro selected ribozymes. *Nature* **390**: 96–100; Unrau PJ, Bartel DP. (1998) RNA-catalysed nucleotide synthesis. *Nature* **395**: 260–263; Bartel DP, Unrau PJ. (1999) Constructing an RNA world. *Trends in Cell Biol* **9**: M9–M13; Lee N, Bessho Y,

Wei K, *et al.* (2000) Ribozyme-catalyzed tRNA aminoacylation. *Nature Structural Biol* **7:** 28–33; Johnston WK, Unrau PJ, Lawrence MS, *et al.* (2001) RNA-catalyzed RNA polymerization: Accurate and general RNA-templated primer extension. *Science* **292:** 1319–1325; Joyce GF. (2002) The antiquity of RNA-based evolution. *Nature* **418:** 214–221; Lincoln TA, Joyce GF. (2009) Self-sustained replication of an RNA enzyme. *Science* **323:** 1229–1232; Ferretti AC, Joyce GF. (2013) Kinetic properties of an RNA enzyme that undergoes self-sustained exponential amplification. *Biochemistry* **52:** 1227–1235.

25. Ferris JP. (2006) Montmorillonite-catalysed formation of RNA oligomers: The possible role of catalysis in the origins of life. *Philos Trans R Soc Lond B* **361:** 1777–1786.

26. Costanzo G, Pino S, Ciciriello F, Di Mauro E. (2009) Generation of long RNA chains in water. *J Biol Chem* **284:** 33206–33216.

27. Meinert C, Myrgorodska I, de Marcellus P, *et al.* (2016) Ribose and related sugars from ultraviolet irradiation of interstellar ice analogs. *Science* **352:** 208–212.

28. Ricardo A, Carrigan MA, Olcott AN, Benner SA. (2004) Borate minerals stabilize ribose. *Science* **303:** 196.

29. Shapiro R. (1999) Prebiotic cytosine synthesis: A critical analysis and implications for the origin of life. *Proc Natl Acad Sci USA* **96:** 4396–4401.

30. Fuller WD, Sanchez RA, Orgel LE. (1972) Studies in prebiotic synthesis VI. Synthesis of purine nucleosides. *J Mol Biol* **67:** 25–33; Fuller WD, Sanchez RA, Orgel LE. (1972) Studies in prebiotic synthesis. VII Solid-state synthesis of purine nucleosides. *J Mol Evol* **1:** 249–257; Orgel LE. (2004) Prebiotic chemistry and the origin of the RNA world. *Crit Rev Biochem Mol Biol* **39:** 99–123.

31. Sutherland's method synthesized cytosine ribonucleotide by the following steps. Cyanamide was reacted with glycolaldehyde to obtain 2-amino-oxazole. The latter was added to glyceraldehyde to yield the pentose amino-oxazolines including the arabinose derivative. The arabinose amino-oxazoline was then reacted with cyanoacetylene to give the anhydroarabinonucleoside, which was then phosphorylated to yield ß-ribocytidine-2′,3′-cyclic phosphate, an activated ribonucleotide needed for RNA synthesis. The corresponding uracil nucleotide was subsequently converted from the cytosine nucleotide by ultraviolet irradiation. See: Powner MW, Gerland B, Sutherland JD. (2009)

Synthesis of activated pyrimidine ribonucleotides in prebiotically plausible conditions. *Nature* **459:** 239–242.

32. Benner S. (2009) *Life, the Universe and the Scientific Method.* The FfAME Press, Gainesville, Florida.
33. Joyce GF. (1989) RNA evolution and the origins of life. *Nature* **338:** 217–224.
34. Cairn-Smith AG. (1966) The origin of life and the nature of the primitive gene. *J Theoretical Biol* **10:** 53–88; Cairns-Smith AG. (1982) *Genetic Takeover and the Mineral Origins of Life.* Cambridge Univ. Press, Cambridge, UK; Cairns-Smith AG. (1985) *Seven Clues to the Origin of Life.* Cambridge Univ. Press, Cambridge, UK.
35. Pitsch S, Wendeborn S, Jaun B, Eschenmoser A. (1993) Why pentose and not hexose nucleic acids? Pyranosyl RNA (p-RNA). *Helv Chim Acta* **76:** 2161–2183.
36. Schöning K, Scholz P, Guntha S, et al. (2000) Chemical etiology of nucleic acid structure: the α-threofuranosyl-(3'->2') oligonucleotide system. *Science* **290:** 1347–1351.
37. Chaput JC, Switzer C. (2000) Nonenzymatic oligomerization on templates containing phosphoester-linked acyclic glycerol nucleic acid analogues. *J Mol Evol* **51:** 464–470.
38. Nielsen PE, Egholm M, Berg RH, Buchardt O. (1991) Sequence-selective recognition of DNA by strand displacement with a thymine-substituted polyamide. *Science* **254:** 1497–1500.
39. Joyce GF. (2002) The antiquity of RNA-based evolution. *Nature* **418:** 214–221.
40. Crick F, Foreword to Gesteland RF, Cech TR, Atkins JF. eds. (2006) *The RNA World.* 3rd ed. Cold Spring Harbor Lab. Press.; Crick FH, Orgel LE. (1973) Directed Panspermia. *Icarus* **19:** 341–348; Crick F. (1981) *Life Itself.* MacDonald, London.
41. Bartel DP, Szostak JW. (1993) Isolation of new ribozymes from a large pool of random sequences. *Science* **261:** 1411–1418.
42. Yarus M. (2010) *Life from an RNA World: The Ancestor Within.* Harvard Univ. Press, Cambridge, MA, p. 180.
43. Cech TR, Moras D, Nagai K, Williamson JR. (2006) The RNP World. In Gesteland RF, Cech TR, Atkins JF. eds. *The RNA World.* 3rd ed. Cold Spring Harbor Lab. Press.

44. Shapiro R. (1986) *Origins: A Skeptic's Guide to the Creation of Life on Earth.* Summit Books, New York.
45. Mansy SS, Szostak JW. (2008) Thermostability of model protocell membranes. *Proc Natl Acad Sci USA* **105**: 13351–13355; Zhu TF, Szostak JW. (2009) Coupled growth and division of model protocell membranes. *J Am Chem Soc* **131**: 5705–5713.
46. Wachtershauser G. (1992) Groundworks for an evolutionary biochemistry: The iron-sulfur world. *Progress Biophys and Molec Biol* **58**: 85–201.
47. Wachtershauser G. (2006) From volcanic origins of chemoautotrophic life to Bacteria, Archaea and Eukarya. *Philos Trans R Soc Lond B* **361**: 1787–806.
48. Martin W, Baross J, Kelley D, Russell MJ. (2008) Hydrothermal vents and the origin of life. *Nature Rev Microbiol* **6**: 805–814.
49. Ferris JP, Hill AR Jr, Liu R, OrgelL E. (1996) Synthesis of long prebiotic oligomers on mineral surfaces. *Nature* **381**: 59–61.
50. Hanczyc MM, Fujikawa SM, Szostak JW. (2003) Experimental models of primitive cellular compartments: Encapsulation, growth and division. *Science* **302**: 618–622.
51. Saladino R, Crestini C, Ciciriello F, *et al.* (2009) From formamide to RNA: the roles of formamide and water in the evolution of chemical information. *Res in Microbiol* **160**: 441–448; Saladino R, Crestini C, Pino S, *et al.* (2012) Formamide and the origin of life. *Physics of Life Reviews* **9**: 84–104; Senanayake SD, Idriss H. (2006) Photocatalysis and the origin of life: synthesis of nucleoside bases from formamide on TiO2 (001) single surfaces. *Proc Natl Acad Sci USA* **103**: 1194–1198.
52. Copley SD, Smith E, Morowitz HJ. (2007) The origin of RNA world: Co-evolution of genes and metabolism. *Bioorganic Chem* **35**: 430–443.
53. Calvin M. (1969) *Chemical Evolution.* Oxford Univ. Press, New York.
54. de Duve especially emphasized the "high-energy" molecules such as the pyrophosphates and thioesters. See: de Duve C. (2013) The other revolution in the life sciences (Lett. to editor). *Science* **339**: 1148; de Duve C. (1995) *Vital Dust: Life as a Cosmic Imperative.* Basic Books, New York; de Duve C. (2002) *Life Evolving.* Oxford Univ. Press, New York.
55. Patel BH, Percivalle C, Ritson DJ, *et al.* (2015) Common origins of RNA, protein and lipid precursors in a cyanosulfidic protometabolism. *Nature Chem* **7**: 301–307.

56. Bar-Nun A, Kochavi E, Bar-Nun S. (1994) Assemblies of free amino acids as possible prebiotic catalysts. *J Mol Evol* **39:** 116–122.
57. Griffith EC, Vaida V. (2012) In situ observation of peptide bond formation at the water-air interface. *Proc Natl Acad Sci USA* **109:** 15697–15701.
58. Mulkidjanian AY, Bychkov AY, Dibrova DV, *et al.* (2012) Origin of first cells at terrestrial, anoxic geothermal fields. *Proc Natl Acad Sci USA* **109:** E821–E830.
59. Yarus M. (2010) *Life from an RNA World: The Ancestor Within.* Harvard Univ. Press, Cambridge, MA, p. 163.
60. Taylor FJ, Coates D. (1989) The code within the codons. *BioSystems* **22:** 177–187.
61. Eigen M, Winkler-Oswatitsch R. (1981) *Naturwissenschaften,* **68:** 282–292.
62. Yarus M, Widmann JJ, Knight R. (2009) RNA-amino acid binding: a stereo chemical era for the genetic code. *J Mol Evol* **69:** 406–429.
63. Yarus M. (2010) *Life from an RNA World: The Ancestor Within.* Harvard Univ. Press, Cambridge, MA, p. 172.
64. de Pouplana LR, Turner RJ, Steer BA, Schimmel P. (1998) Genetic code origins: tRNAs older than their synthetases? *Proc Natl Acad Sci USA* **95:** 11295–11300.
65. Erives A. (2011) A model of proto-anti-codon RNA enzymes requiring L-amino acid homochirality. *J Mol Evol* **73:** 10–22.
66. Morowitz HJ. (2002) *The Emergence of Everything.* Oxford Univ. Press, New York.
67. Kauffman S. (2004) The autonomous agents. In Barrow JD, Davies PCW, Harper CL Jr. eds. *Science and Ultimate Reality.* Cambridge Univ. Press, New York.
68. Kauffman S. (1995) *At Home in the Universe.* Oxford Univ. Press, New York; Daley AJ, Girvin A, Kauffman SA, *et al.* (2002) Simulation of a chemical autonomous agent. *Zeits Phys Chem* **216:**41.
69. Waldrop MM. (1992) *Complexity: The Emerging Science at the Edge of Order and Chaos.* Simon & Schuster, New York. (Quoted with permission from Simon & Schuster, and Sterling Lord Literistic.)
70. Ke Y, Ong LL, Shih WM, Yin P. (2012) Three-dimensional structures self-assembled from DNA bricks. *Science* **338:** 1177–1183; Gothelf KV. (2012) LEGO-like DNA structures. *Science* **338:** 1159–1161.

71. Crick F. (1981) *Life Itself: Its Origin and Nature.* Simon & Schuster, New York, p. 88.
72. Kauffman S. (1995) *At Home in the Universe.* Oxford Univ. Press, New York, p. 31.
73. Orgel LE. (October 1994) The origin of life on the earth. *Scientific Am* **271**: 78.
74. Peckham M. (1959) *The Origin of Species: A Variorum Text.* Univ. of Pennsylvania Press, Philadelphia, p. 748.
75. Monod stated: "What, before the event (appearance of life on Earth), were the chances that this would occur? The present structure of the biosphere far from excludes the possibility that the decisive event occurred only once. Which would mean its a priori probability was virtually zero…This idea is distasteful to many scientists. Science can neither say nor do anything about a unique occurrence…If it was unique, as may perhaps have been the appearance of life itself, then before it did appear its chances of doing so were infinitely slender. The universe was not pregnant with life nor the biosphere with man. Our number came up in the Monte Carlo game. " See: Monod J. (1970) *Le Hazard et la Necessite.* (Editions de seuil, Paris), trans. Wainhouse AG. (1971) *Chance and Necessity.* Knopf, New York, pp. 144–145. (Quoted with permission from Penguin Random House.)
76. Eigen M. (1992) *Steps Towards Life.* Oxford Univ. Press, Oxford.
77. The simplest free-living organism we can find today is the bacteria *Mycoplasma genitalium*, which has 525 genes. See: Fraser CM, Gocayne JD, White O, *et al.* (1995) The minimum gene complement of *Mycoplasma genitalium. Science* **270**: 397–403. In a different approach, Venter and his colleagues synthesized the genome of *Mycoplasma mycoides* and determined the minimum number of genes necessary for survival and reproduction. They ended up with 473 genes. See: Hutchison CA III, Chuang R-Y, Noskov VN, *et al.* (2016) Design and synthesis of a minimal bacterial genome. *Science* **351**: aad6253. (DOI: 10.1126/science.aad6253.) It should be noted that these bacteria with a synthetic genome thrive only under the most ideal laboratory conditions, with all the nutrients provided for in the culture medium. The theoretical first living organism on Earth was not so fortunate, and would presumably need many more genes to cope with the harsh prebiotic environment.

78. In a letter to Wallich in 1882, Darwin wrote: "I had intentionally left the question of the Origin of Life uncanvassed as being altogether *ultra vires* in the present state of our knowledge, and that I dealt only with the manner of succession." See: de Beer G. (1959) *Some Unpublished Letters of Charles Darwin*. Notes Rec. R. Soc. Lond., **14:** 12–66.
79. Richards Dawkins, for one, seemed to entertain this possibility in his book: "The present lack of a definitely acceptable account of the origin of life should certainly not be taken as a stumbling block for the whole Darwinian world view." See: Dawkins R. (1986) *The Blind Watchmaker*. Norton, New York, (1996 paperback ed. p. 166).
80. In an article published in the Wall Street Journal (September 12, 2009), Dawkins put it more bluntly, "Evolution is the creator of life."
81. Cited by Dawkins in his book. See: Dawkins R. (1986) *The Blind Watchmaker*. Norton, New York, (1996 paperback ed. p. 131).
82. Spiegelman S. (1967) An in vitro analysis of a replicating molecule. *Am Scientist* **55:** 63–68; Mills DR, Peterson RI, Spiegelman S. (1967) An extracellular Darwinian experiment with a self-duplicating nucleic acid molecule. *Proc Natl Acad Sci USA* **58:** 217–224; Kacian DL, Mills DR, Kramer FR, Spiegelman S. (1972) A replicating RNA molecule suitable for a detailed analysis of extracellular evolution and replication. *Proc Natl Acad Sci USA* **69:** 3038–3042.
83. Darwin C. (1859) *The Origin of Species*. last sentence.
84. Dawkins R. (1976) *The Selfish Gene*. Oxford Univ. Press, Oxford; 1989 edition.
85. Dawkins R. (1982) *The Extended Phenotype*: The Gene as the Unit of Selection. Oxford: Freeman, p. 114.
86. Small segments of RNA can regulate gene expression by acting either directly on DNA or indirectly on messenger RNA.
87. Jablonka E, Lamb MJ. (2005) *Evolution in Four Dimensions*. MIT Press, Cambridge, MA.
88. Wickner RB. (1994) URE3 as an altered URE2 protein: evidence for a prion analog in *Saccharomyces cerevisiae*. *Science* **264:** 566–569; Sondheimer N, Lindquist S. (2000) Rnq1: an epigenetic modifier of protein function in yeast. *Molecular Cell* **5:** 163–172; Li RM, Lindquist S. (2000) Creating a protein-based element of inheritance. *Science* **287:** 661–664.

89. Cannon WB. (1932) *The Wisdom of the Body.* Norton, New York. The title of the book was borrowed from an earlier lecture given by E.H. Starling.
90. Kuhn TS. (1970) *The Structure of Scientific Revolutions.* (Foundations of the Unity of Science, Vol. II, No. 2), 2nd ed. Univ. Chicago Press, Chicago.
91. Russell B. (1962) *Essays in Skepticism.* Wisdom Library (A Division of The Philosophical Library), New York, p. 83.
92. Polanyi M. (1968) Life's irreducible structure. *Science* **160:** 1308–1312.
93. Ernst Mayr stated that biology distinguishes itself from physics by being information-rich. Information, of course, is the outcome of organization. See: Mayr E. (2004) *What Makes Biology Unique?* Cambridge Univ. Press, Cambridge, UK.
94. Hofstadter stresses pattern formation and the recursive, self-referential properties of life, but does not provide a solution as to how this came about. See: Hofstadter DR. (1979) *Godel, Escher, Bach: An Eternal Golden Braid.* Basic Books, New York.
95. Wolpert L, Richards A. (1988) *A Passion for Science.* Oxford Univ. Press, Oxford, p. 37; Rees MJ. (2011) Back to the Beginning. *Newsweek Magazine,* Dec. 26, p. 51–52; Rees MJ. (2012) *From Here to Infinity.* Norton, New York. (Quoted with permission from Martin J. Rees.)
96. The analogy is like looking at the hardware of a computer and trying to predict how many programs can be written to run on it.

Chapter 4

The Microbial *Self*

If amoebas were as large as whales, would you get out of their way?

— H.S. Jennings

Overview: *Unicellular organisms express a sense of self, as manifested by their strong tendency to survive and procreate. Examples can be seen in prokaryotes and simple eukaryotes (cells with a nucleus) in their feeding behavior, avoidance of danger and defense of invasion. Some of their behaviors are quite complex. The unusual ability of slime mold to change from a unicellular to a multicellular existence and vice versa is a prototype of a flexible self.*

Microbes are unicellular organisms that range from the barely visible to those invisible to the naked eye. They can replicate endlessly and rapidly, doubling in less than an hour. They have a simple genome that can mutate rapidly to adapt to a changing environment. They can extract energy from varied sources for survival. All these properties make them evolutionarily the most successful (in terms of number) life forms on Earth. Together they comprise the largest biomass on the globe — about 80%. These unseen organisms inhabit every millimeter of land and sea, and are also found, though in lesser number, in the air. They live on the highest peak of Mount Everest, thirty thousand feet above sea level. They survive under high temperatures in thermal geysers and under enormous hydrostatic pressure on the deepest ocean floor. Further below the sea floor, they go as deep as 2.5 kilometers. Within our body, bacteria cohabit with human cells, mainly in the gut, outnumbering our own by a factor of ten. This chapter is dedicated to showing that, even in the microbial world, *self* is well in place and clearly expressed.

Microorganisms are classified into bacteria, archaea, and protists. The first two are prokaryotes (having no nucleus), whereas the third are eukaryotes (with a nucleus). Despite the name, archaea are not the oldest in evolutionary history; bacteria are. Prokaryotes do not possess well-formed structures within the cell. They have a circular DNA and only one set of genes (haploid). They reproduce asexually by binary fission. Nonetheless, genetic mixing is frequent through transfer of DNA across individuals, resulting in a flexible gene pool. Archaea differ from bacteria in two major respects: the genome of archaea contains introns as well as exons; their DNA is bound to a special protein called histone. (Introns are portions of the DNA chain that do not encode protein.) These features put them closer to eukaryotes than bacteria. Unlike bacteria, which are widely distributed, archaea are restricted to thrive in more extreme conditions, such as high temperature, high pressure, and high acidity. Hence they are frequently associated with the term "extremophiles."

Unlike the prokaryotes, eukaryotes have two sets of genes (diploid) that can undergo meiosis (splitting into haploids) during reproduction. Protists are unicellular eukaryotes that are neither animals nor plants. They are larger than bacteria and archaea. Protists are classified into protozoans, slime molds and algae. Examples of protozoans are amoebas, paramecia, and stentors. Slime molds are unique in that they are both unicellular and multicellular depending on food supply. Algae contain chloroplasts and are capable of photosynthesis.[1]

4.1 Bacterial Expression of *Self*

Despite being the simplest free-living organisms, bacteria do express a rudimentary *self*, as can be seen in the following examples:

4.1.1 *Defense against adverse environment*

When the environmental condition is unfavorable, some bacteria can protect themselves by forming a spore. They produce a thick coat to surround the internal structures and can stay dormant in this manner for extended periods of time, even thousands of years, unharmed by harsh treatments including boiling, radiation, desiccation, and chemical disinfectants. Once they find themselves in a suitable location, they will revive and lead a normal life again.

4.1.2 *Defense against invading viruses*

Bacteria constantly face invasion by viruses. One way to stave off these foreign DNAs is to destroy the viral strands with a type of enzyme called "restriction enzymes," like cutting a ribbon with a pair of scissors. These are enzymes targeted against certain nucleotide sequences that are frequently found in viruses. It is a type of innate immunity since the enzymes are there whether or not the viruses are present. The bacteria's own DNA is protected by methylation and therefore is not cleaved by the enzymes. In this instance, the microbes are able to distinguish *self*-DNA from non-*self*-DNA.

Bacteria can also mount an acquired immune defense against viruses. They can incorporate tiny fragments of an invading virus into their genome, serving as a historical record of the invasion — a sort of biochemical memory. The inserted foreign pieces of DNA can generate small RNAs corresponding to the sequence of the virus. Upon reinfection, these RNAs (called CRISPR-RNAs) can recognize the DNA sequence of the invaders and bring about their destruction by an associated enzyme. This is an adaptive immunity directed specifically against the particular invader.[2]

4.1.3 *Transfer of DNA and silencing of harmful foreign DNA*

A bacterium possesses chromosomal DNA in the form of a single long circular chain. This DNA is involved in cell division and is the main genome of the bacterium. In addition, it may have one or more short DNAs (called plasmids), that are not involved in division but can readily be transferred from one bacterium to another through cell contact, a process called conjugation, the equivalent of mating in higher organisms. Sometimes a fragment of the chromosomal DNA can also be transferred between distantly related organisms. Both types of horizontal transfer of genetic material usually benefit the recipient and play a role in evolution.[3] However, bacteria can reject an acquired DNA if it is harmful. Some bacteria produce a protein that can distinguish *self-* from non-*self-*DNA by its chemical composition, such as the ratio of the GC/AT (guanine-cytosine/adenine-thymine) nucleotides.[4] Other bacteria can eliminate unwanted foreign plasmids by cleaving them with an enzyme.[5]

4.1.4 Social recognition in a community

Some bacteria form swarming colonies in culture. Colonies of the same strain coalesce to form a single large colony, but those of different strains remain separated, an example of distinguishing *self* from non-*self*. A gene responsible for this recognition has been identified.[6]

4.1.5 Quorum sensing for mass action

Quorum sensing is the way bacteria communicate with one another using chemical messages. It allows the bacteria to estimate the population density, to determine whether they have reached a critical mass, and to carry out processes that requires collective action to be effective, such as the massive secretion of a toxin to overpower the immune system of the host. Other collective actions include biofilm formation and sporulation. Biofilms are community-level stress response in which a network of channels is formed that enhances transport of nutrients.[7] Bacteria use multiple chemical signals to "talk" to others of the same species, but they also use them for across-species interaction.[8] Some bacteria can engage in electron transfer on their outer membrane from one cell to another, forming long chains of information flow. In this manner thousands of microbes can act in concert as a multicellular unit. This phenomenon results in sharing of an energy source and aids in the effectiveness of quorum sensing.[9]

4.1.6 *Bacterial individuality*

Motile bacteria engage in "random walk" (by tumbling) in search of nutrients. When they detect a positive gradient in the environment, they stop tumbling and swim in the direction of the food. But once the gradient disappears, random walk resumes. Even if genetically identical, some bacteria appear to be more "nervous" and tumble more often, while others do not jitter as much. The tumbling frequency is correlated with the sensitivity to an attractant and differs from cell to cell.[10]

4.2 Protists: The One-cell Eukaryotes

In contrast to prokaryotes, eukaryotes possess well-formed structures inside the cell called organelles. These include the nucleus, mitochondria, lysosomes and Golgi apparatus. Compared to prokaryotes, their strategy for survival and preservation of *self* is much more sophisticated. They can literally swallow other organisms alive by phagocytosis (cell eating) and take in liquid material by pinocytosis (cell drinking). Even for the simplest form of eukaryotes, the protists, a plethora of behavior is developed to interact with the environment and for the procurement of food. Here we shall look at the behavior of some protists.

4.2.1 *Amoeba feeding behavior*

Amoebas are unicellular organisms that live at the bottom of ponds or other moist places including soil. A typical amoeba is about 0.3 mm in size. It consists of a cell membrane that surrounds a fluid-like cytoplasm made up of an outer thin ectoplasm and a bulky endoplasm. The amoeba moves along a solid surface by a rolling motion of the cell body and the extension and retraction of finger-like projections of the cytoplasm called pseudopodia. They ingest solid food by surrounding it with the pseudopodia and incorporating it into a round structure called food vacuole, which contains the digestive enzymes. After absorption, the residue in the food vacuole is extruded and left behind in the track of the organism. Ingestion and egestion take place in any part of the organism, as there is no specialized area of the cell that can be identified as mouth or anus. Other than the nucleus, the other prominent structures in the cytoplasm are the contractile vacuoles that contain water. Water enters the cytoplasm by osmosis because the osmotic pressure inside the cell is higher than that in the surroundings. The amoeba expels the accumulated water by periodic contraction of the contractile vacuoles.

Amoebas take up as food other microorganisms such as algae, diatoms, bacteria and other protozoans. They can distinguish food from inedible materials and use different tactics in approaching different food. For such a simple organism, the way they achieve their goal is quite remarkable. The following example shows how an amoeba engulfs a Euglena cyst (Fig. 4.1). The cyst is spherical and easily rolls away as soon as the approaching amoeba touches it. After pursuing in vain for a while, the amoeba changes tactics by stopping the forward movement. Instead, one pseudopodium is sent out from each side to surround the cyst. At the same time the cytoplasm moves over the immobilized cyst to cover it. In this manner the amoeba easily swallows its prey.[11]

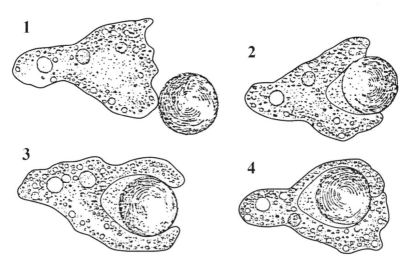

Fig. 4.1. An amoeba engulfing a euglena cyst. Numbers indicate sequence of events. [See Note 11; permission Columbia Univ. Press.]

4.2.2 Stentor avoidance behavior

Stentors are unicellular aquatic organisms measuring about 1 mm in length. They have a trumpet-shaped body and normally attach their narrow end to the vegetative debris in a pond. The open end is the oral cavity, which actively sucks in food by means of a vortex created by a row of cilia surrounding the cavity. The food passes into an oral pouch then into the gullet before turning into food vacuoles. The stentor will stay in one place to feed until food is deprived, or until the condition is no longer favorable, whereupon it swims away to another location. Stentors exhibit a high degree of food selection. At the oral pouch living objects are taken in, whereas indigestible particles are rejected. Among living organisms some species (such as euglena) are preferred to others (such as chilomonas) as food.

The most elaborate behavior of stentors is one of avoidance of a noxious stimulus (Fig. 4.2), as documented by H. S. Jennings.[11] Jennings did an experiment wherein a cloud of indigestible particle such as

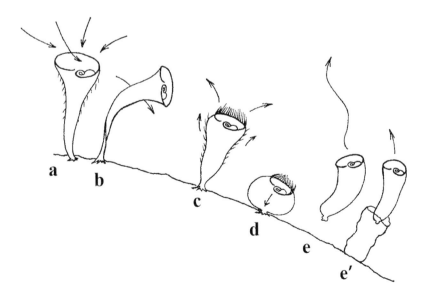

Fig. 4.2. A stentor responding to carmine particles. Letters indicate sequence of events. See text for details. [See Note 12; permission Elsevier Books.]

carmine (a pigmented substance) was continuously poured toward the oral cavity of a stentor. Following an initial period of indifference, the stentor first bends its body to avoid the particles. When this response fails to remove the undesirable stimulus, the stentor abruptly reverses the beat of the cilia in an attempt to propel the particles away. If the stimulus persists, the stentor suddenly contracts, and then slowly resumes its normal shape. After about fifteen minutes of such contraction and relaxation cycles, and in the continuing presence of the noxious particles, the stentor violently sets itself free from the adhered surface and swims to a new location. On landing, the stentor first explores the new surface by crawling with the mouth touching the surface. Once a suitable spot is found, it secretes a gelatinous material to firmly glue itself to the new surface. This repertoire of behavior was observed repeatedly in many individuals, with only minor variations.

Note that this is not a series of rigid, stereotyped behaviors started off by a given stimulus. The organism expresses a goal to eliminate the unwelcome change in environment by using various responses, ranging from the mildest to the most drastic measure. Once the goal is achieved, the stentor will go no further. It is only when one trick fails to remove the noxious stimulus that the organism attempts the next stronger response, ending with detachment from the site and swimming away to a new location. That the stentor exerts an apparent "attempt" to overcome the obstacle is obvious to the observer.

The sequence of responses can be interpreted in more detail as follows. First, the stimulus (carmine particles), though not causing immediate injury, must be perceived as annoying and in the long run harmful, and therefore should be avoided or removed. The reactions are of a graded nature, from the most effortless to the most forceful. At first, the stentor ignores the annoyance by continuing its normal activity. Then a slight effort is exerted by turning the mouth away, while still engaged in feeding. When this is ineffective in avoiding the stimulus, the feeding stops temporarily and the ciliary current is reversed to get rid of the particles. If this attempt still fails, the normal activity is completely interrupted, and a strong contraction of the body ensues, in an effort to remove the noxious stimulus while still remaining in the same place, ready to resume its normal activity any moment. But if this is unsuccessful after many tries, the last resort appears, which is to abandon the feeding place and swim away to a new location no longer threatened by the stimulus, a much more drastic effort than all the preceding steps. It should be noted that the nature and intensity of the stimulus remains unchanged throughout the experiment. The reactions are such that one step does not automatically lead to the next. At each stage the stentor needs to evaluate the outcome of the previous steps. Only if the preceding steps are ineffective will a new response be brought into play. The chain of reactions terminates at any point as soon as the stimulus is removed. These organisms exhibit a rudimentary short-term memory, as each new response is modified by the earlier ones.[11,12]

4.2.3 *Slime molds: A link between unicellularity and multicellularity*

Slime molds are unique among protists in that they are capable of switching between single-cell and multi-cell existence. The species *Dictyostelium discoideum* normally lives in soil and moist leaf litter, feeding on bacteria and yeast cells. When food supply is abundant, they live as independent amoebas and divide every few hours. When food is in short supply, they undergo a series of changes to ensure their survival (Fig. 4.3). First, individual cells start to synthesize and emit a chemo-attractant, cyclic AMP, in a random manner. Next, the cells aggregate into several local centers with synchronized cyclic AMP pulses. Eventually one center dominates in pulsing, into which all others coalesce to form a single aggregate. The cells are held together by adhesion molecules, resulting in a "slug" of about 2-4 mm long. Each slug behaves as one organism and glides on the ground, leaving behind a track of slime (hence the name).Once arriving in a favorable spot, the slug stops moving and differentiates into a fruiting body,

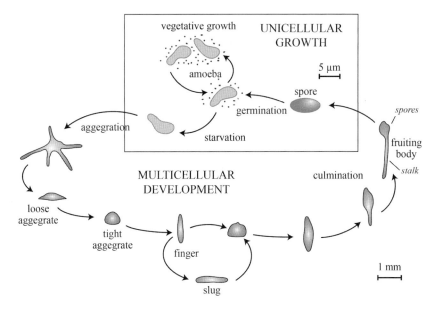

Fig. 4.3. Aggregation of the slime mold *Dictyostelium discoideum*. The top part shows the free-living, individual cells. The left and lower parts show the grouping of the cells into a moving slug when food supply is low. The right part depicts the formation of a fruiting body whereby the spores germinate into single cells. [Wikimedia Commons.]

consisting of a stalk and spores. The mature spores are released and, when conditions are favorable, germinate into free-living, single-cell amoebas to start the life cycle again. The entire process takes about thirty hours. This clever trick for self-preservation is not observed in other unicellular eukaryotes.[13]

4.2.4 *Self-optimization of the slime mold Physarum polycephalum*

The *Physarum* is a cell with multiple nuclei. In the growing phase, the cell increases in size and the nuclei multiply, but the cell does not divide. All the nuclei are suspended in a contiguous sheet of cytoplasm that covers an area like a pancake. When nutrients are limited, the cell forms an interconnected network of cytoplasm. The network is highly dynamic, with cytoplasm continuously flowing from one part to another. In this manner the cell crawls slowly from one place to another in search of food. If food is scattered in several locations, as in the case of oat flakes placed on a surface, the cellular network positions itself within hours to maximize contact with the food sources. To do this the cell connects the multiple food supplies along the shortest possible path, even finding the best way through a maze. Surprisingly, the networking simulates a purposefully designed mathematical model for a railway system. The mechanism for this self-optimization algorithm is yet to be explained.[14] *Physarum* slime mold is also capable of forming spatial "memory" in navigating a U-shape barrier to reach a chemo-attractive goal even in the absence of a gradient guidance (Fig. 4.4). It turns out that as the slime mold forages, it avoids areas that contain extracellular slime, which the mold leaves behind as it moves. It uses a chemical feedback as "memory" of the past.[15]

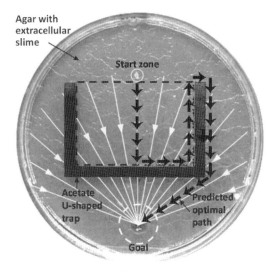

Fig. 4.4. Navigation of *Physarum polycephalum* around a U-shaped trap. The multiple short arrows indicate the path taken by the organism to reach the goal (food source). [See Note 15; courtesy US Natl. Acad. Sci.]

4.3 A Bigger Question

What is most thought provoking about microbes is the sheer size of these puny beings. Take bacteria for instance. A million of them can barely make up a visible dot on this page. We humans are several trillion times their body size, and our brain functions with 100 billion (1×10^{11}) interconnecting neurons. Compare this to an amoeba or stentor, which does not even have a nerve fiber and an excitable membrane to propagate electrical impulse. Yet with such paltry equipment, the microbes, like us, are capable of carrying out some of the most critical functions of life — the expression of *self* through food seeking and avoidance of danger. The question is, do microbes possess a rudimentary form of awareness and willfulness? H.S. Jennings, a noted biologist, gave the following perplexing statement:

> "We do not usually attribute consciousness to a stone, because this would not assist us in understanding or controlling the behavior of the stone. ……On the other hand, we usually attribute consciousness to the dog, because this is useful; it enables us practically to appreciate, foresee, and control its actions much more readily than we could otherwise do so. If Amoeba were so large as to come within our everyday ken, I believe it beyond question that we should find similar attribution to it of certain states of consciousness a practical assistance in foreseeing and controlling its behavior. Amoeba is a beast of prey, and gives the impression of being controlled by the same elemental impulses as higher beasts of prey. If it were as large as a whale, it is quite conceivable that occasions might arise when the attribution to it of the elemental states of consciousness might save the unsophisticated human being from the destruction that would result from the lack of such attribution. In such a case, then, the attribution of consciousness would be satisfactory and useful. …… But such impressions and suggestions of course do not demonstrate the existence of consciousness in lower organisms. Any belief on this matter can be held without conflict with the objective facts. All that experiment and observation can do is to show us whether the behavior of lower organisms is objectively similar to the behavior that in man is accompanied by consciousness. If this

question is answered in the affirmative, as the facts seem to require, and if we further hold, as is commonly held, that man and the lower organisms are subdivisions of the same substance, then it may perhaps be said that objective investigation is as favorable to the view of the general distribution of consciousness throughout animals as it could well be. But the problem as to the actual existence of consciousness outside of the self is an indeterminate one; no increase of objective knowledge can ever solve it. Opinions on this subject must then be largely dominated by general philosophical considerations, drawn from other fields."[16]

Did Jennings go too far in anthropomorphizing these tiny creatures? I have no clear-cut answer, but I shall touch on this issue in *Chapter 13: Self from Within: The Introspective Self*. Should you need more food for thought, please look at the following statement from a prominent neuro-anatomist, Ramon y Cajal, as quoted by an equally prominent neurophysiologist, Charles Sherrington: "I remember that once I spent twenty hours continuously at the microscope watching the movements of a sluggish leukocyte in its *laborious efforts* (emphasis mine) to escape from a blood capillary."[17] Leukocytes are animal white blood cells, but their microscopic size and show of apparent "effort" make them behave much like free living microbes.

Notes and References

1. The reader may wonder why I do not include viruses in the topic of microbial self. The reason is that whether or not viruses are living things is controversial, as they do not carry on an independent subsistence. They "hijack" the cellular machinery of other living things to survive.
2. Horvath P, Barrangou R. (2010) CRISPR/Cas, The immune system of bacteria and archaea. *Science* **327:** 167–170.
3. Ochman H, Lawrence JG, Groisman EA. (2000) Lateral gene transfer and the nature of bacterial innovation. *Nature* **405:** 299–304.
4. Navarre WW, Porwollik S, Wang Y, *et al.* (2006) Selective silencing of foreign DNA with low GC content by the H-NS protein in salmonella. *Science* **313:** 236–238.

5. Jorg V. (2014) A bacterial seek-and-destroy system for foreign DNA. *Science* **344:** 972–973.
6. Gibbs KA, Urbanowski ML, Greenberg EP. (2008) Genetic determinants of self identity and social recognition in bacteria. *Science* **321:** 256–259.
7. Wilking JN, Zaburdaev V, De Volder M, *et al.* (2013) Liquid transport facilitated by channels in Bacillus subtilis biofilms. *Proc Natl Acad Sci USA* **110:** 848–852.
8. Ng WL, Bassler BL. (2009) Bacterial quorum-sensing network architectures. *Ann Rev Genet* **43:** 197–222.
9. Perbadian S, El-Nagger MY. (2012) Multistep hopping and extracellular charge transfer in microbial redox chains. *Phys Chem Chem Phys* **14:** 13802–13808.
10. Spudich JL, Koshland DE Jr. (1976) Non-genetic individuality: Chance in the single cell. *Nature* **262:** 467–471.
11. Jennings HS. (1906) *Behavior of the Lower Organisms.* Columbia Univ. Press, New York.
12. Tartar V (1961) *The Biology of Stentor.* Pergamon Press, New York.
13. Mato JM, Losada A, Nanjundiah V, Konijn TM.(1975) Signal input for a chemotactic response in the cellular slime mold Dictyostelium discoideum. *Proc Nat Acad Sci USA* **72:** 4991–4993; Gregor T, Fujimoto K, Masaki N, Sawai S. (2010) The onset of collective behavior in social amoeba. *Science* **328:** 1021–1025.
14. Tero A, Takagi S, Saigusa T, *et al.* (2010) Rules for biologically inspired adaptive network design. *Science* **327:** 439–442; Marwan W (2010) Amoeba-inspired network design. *Science* **327:** 419–420.
15. Reid CR, Latty T, Dussutour A, Beekman M. (2012) Slime mold uses an externalized spatial "memory" to navigate in complex environments. *Proc Natl Acad Sci USA* **109:**17490–17494.
16. Jennings HS. (1906) *Behavior of the Lower Organisms.* Columbia Univ. Press, New York, p. 337. (Quoted with permission from Columbia Univ. Press.)
17. Ramon-y-Cajal S. *Recollections of my Life.* ed. 3, II, p. 171; quoted by Sherrington C. (1940), *Man on His Nature.* Cambridge Univ. Press, Cambridge, UK, Chap. 3.

Chapter 5
The Plant *Self*

I believe a leaf of grass is no less than the journey-work of the stars.

— Walt Whitman

Overview: *Plants have a strong sense of self though less obvious to us. Their roots can distinguish kin from strangers. They "talk" to one another with volatile chemicals and root exudates. They alert their neighbors when attacked by herbivores. They fight against pathogens and even have an innate immune system. Whereas animals roam around in search of friendly habitats, plants readily switch their metabolic strategies to cope with local, unexpected environmental changes. Their way of dealing with animals is subtle and clever. They recruit animals to spread their genes by wrapping their seeds in tasty coatings. They prevent being overeaten by making chemicals that are toxic to animals but harmless to themselves.*

Plants are sessile and spend their life in one place. When kicked, they do not hit back or run away. We humans seldom accord a sense of *self* to these fellow living things. Nevertheless, plants do have a *self* that they carefully maintain and protect, silently. In this chapter I shall bring out some examples of how plants express their *selves*, mainly in three contexts: self-defense, communication, and competition. The three are closely related, as communication is the key mechanism for defense and competition.

5.1 Plant Protection of *Self*

All animals directly or indirectly derive their energy from plants, which in turn obtain their energy from the sun. Thus, plants are under pressure to prevent being over-consumed to extinction. As they cannot turn away from predators, they evolve strategies to ward off intruders. Some plants develop bad taste in their leaves or thorns on their twigs. Others produce compounds poisonous to animals but innocuous to themselves. History tells us that Socrates was executed with hemlock in 399 B.C.E. Hemlock contains a potent alkaloid that causes convulsion, paralysis and death, and was commonly used for execution in ancient Greece. Another historical case was the assassination of the Holy Roman Emperor Charles IV with the Death Cap mushroom (*Amanita phalloides*), which is also responsible for most accidental mushroom poisoning. Among the most poisonous substances is ricin, a protein from the seeds of castor-oil plant (*Ricinus communis*), which is lethal to humans at a dose of 0.5 mg, sized smaller than a grain of salt. Ironically, some plant toxins are fortuitously beneficial to humans as medicines if consumed in moderate amounts, such as the cardiac glycoside digitalis from foxglove. About 25% of all drugs available now are derived from such plant "poisons."

Plants tend to control insects that feed on them. Milkweed leaves are a favorite food for caterpillars of the monarch butterfly. Against this, milkweed develops three lines of defense. First, the leaves are covered with a layer of hair, which must be shaved off by the caterpillar. Second, when bitten, the leaf oozes sticky latex that can drown the insect. Third, the leaf contains a cardiac glycoside that can kill the caterpillar when taken in large amounts.

Another plant strategy is to recruit carnivores that prey on the insects that bite the leaves, through the release of volatile attractants following injury. Lima beans, when infested with spider mites, emit a terpenoid that attracts spider-mite predators to the scene. The same mechanism of defense has been observed in strawberries and other plants. Some plants alter the taste of their leaves when attacked, rendering them less palatable to the herbivores.

The trees in the Serengeti ecosystem of East Africa provide an excellent example of how plants cope with herbivores. First, they cover themselves with spines, needles and hooks. Some trees produce steel-like thorns sharp enough to puncture automobile tires, earning the name of "the puncture tree." The exceedingly painful "Wait-a-bit Acacia" (*Acacia brevispica*) has opposing hooked thorns which cling to passers-by and force them to cry out "wait a bit" or "wait a minute" before disentangling themselves, thus the name. Tall trees like Umbrella Acacia are a favorite food of giraffe, and they protect themselves with thorns on the outer branches where the animals can reach, but the inner branches remain soft and green. Perhaps one of the most "ingenious" ways of defense is adapted by the Whistling Acacia (*Acacia depranolobium*). The tree grows hollow galls the size of golf balls. They also produce flower-like structures (no seeds) that make nectar so sweet that it invites ants to take residence in the galls. The ants' vicious sting drives away any herbivore intruders. The name "whistling" comes from the sound produced when the wind makes its way through the holes of the galls. Lastly, *Acacia robusta* (common name Stinking Acacia) is relatively harmless, but it emits an unpleasant odor that drives away any herbivore that gets close to it.

Plants also protect themselves by spreading out their progeny geographically as far as possible, thus compensating for their lack of mobility. Seeds can be carried by wind, water, and animals (through their fur or digestive tract) to faraway places. Seeds are desiccated embryos that can stand adverse conditions and stay dormant for as long as hundreds of years, only to germinate at the right time in the right place when conditions are favorable. Seeds are well adapted to their environment — some will germinate only after a period of freezing, in order to ensure germination in the spring time and not earlier; others will germinate only after a period of intense heat, in order to ensure germination after a mild forest fire, when abundant fertilizer is available.

Other than biotic stress, plants must face physical (abiotic) hardship as well. Since they cannot run away from adverse environment, they have to develop ways to tolerate and to acclimatize. Physical stress includes extremes of temperature, extremes of humidity, ultraviolet irradiation, nutrition deficiencies, and oxidative stress such as ozone in the air. Plants respond to high temperature by inducing heat-shock proteins, which help to maintain protein conformation, and to low temperature by cold-shock proteins, which protect the RNA machinery. In times of drought, plants tend to increase water absorption from the roots, to increase its entry into the cells, and to minimize water loss from the leaves by closure of the stomata (tiny openings on the underside of the leaves). Oxidative stress due to ozone and ultraviolet radiation is countered by antioxidants, which are abundant in plants. Nutrient deficiency elicits variable biochemical changes aiming at stabilizing the concentration of the particular mineral in question. The stress caused by high soil salinity is responded to by increased intracellular solute particles, including osmotin, a small protein whose intracellular level can reach as high as 12% of the total cell protein.

The interaction of plants and humans is subtle and sophisticated. Over the course of history, mankind has cultivated a number of crops and fruits selected for their taste and/or nutrition value. What usually evades our attention is that the benefit goes both ways. These plants achieve their "goal" of perpetuation by enticing us with their attractive features.

5.2 Extravagant Metabolism as a Strategy for Survival

Both plants and animals need to maintain a relatively constant internal environment (homeostasis) for survival. However, unlike animals that move around to select an optimal location, plants are stuck in one place once their seeds are planted. To cope with fluctuations in external conditions, plants have evolved highly diverse secondary sets of metabolic pathways, many of which are needed only in special situations and not for life maintenance in normal conditions. Hence, overall plant metabolism is many times more complicated than what we can see in an animal. The secondary (non-essential) metabolic pathways in plants arise through permissive genetic mutation, resulting in tremendous chemodiversity and increased adaptability for survival.[1] For example, phenolic and waxy cuticles serve as sunscreens and prevent desiccation. Rubber and lignin provide mechanical support, wound healing, and seed protection. A huge repertoire of compounds is produced not only to cope with changing environment, but also for interspecies interactions that affect pollination. Genomic comparisons suggest that the expansion of plant metabolic pathways occurred 500 million years ago, when plants started to colonize the land, and the process continues ever since.[2]

5.3 Tree Talks and Eavesdropping

Plant-plant communication can be conducted through the air by a group of gaseous substances (Fig. 5.1). About 2,000 of these volatile organic

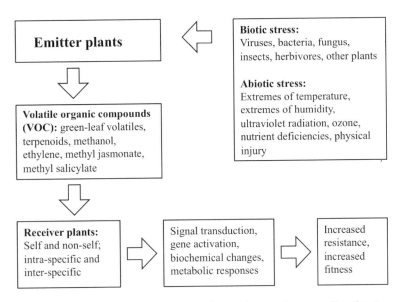

Fig. 5.1. Volatile signals between plants. Plants subject to biotic and/or abiotic stress emit a bouquet of volatile organic compounds (VOC) that is carried in the air to other parts of the same plant (remote from the site of injury) or to other plants of the same or different species. The composition and relative intensity of each component molecule depends on the type of stress and the emitter plant. Upon sensing the stress signal, the receptors in the neighboring plants trigger a series of intracellular signal transduction cascades, leading to gene activation and metabolic alterations. [See Note 6.]

compounds (called VOC) are currently known. These distress signals serve as warning of approaching danger to distant branches of the same tree and to other trees in the vicinity as well. VOCs stimulate receptors in the receiver plants, leading to gene activation and a series of biochemical and metabolic changes, the outcome of which is to increase the resistance to stress before it actually arrives.

It was initially noticed that when a sagebrush plant and a tobacco plant are located next to each other, clipping of the former releases volatile substances that activate the defense readiness of the latter, reducing its susceptibility to herbivory.[3] In a subsequent study using a series of Lima bean plants, it was shown that the receiving plant not only increases its resistance but also releases the same messages, creating relay warning along the line.[4] What is relevant to *self*-recognition is that the danger cues are more effective in eliciting a response between two related plants.[5] Since each plant upon injury emits a mix of chemicals, the composition of which is characteristic of the species, it is possible that plants can decipher the signal and tell whether it comes from a related source. This ability increases biological fitness, as related plants share common natural enemies. Some of the VOCs, such as methyl jasmonate, methyl salicylate, and ethylene, are plant hormones.[6]

The broadcasting of stress signals through the air is a one-way communication and benefits the receivers rather than the emitters. The situation has been compared to kin selection in animals, as the outcome of the defense process is the preservation of the common gene.

5.4 Racing to the Top

Most plants absorb blue-green and red light for photosynthesis. In a dense community, each plant competes with neighboring plants for usable light in a behavior called "shade avoidance syndrome" (SAS). Red light (wave length 655–665 nm) is absorbed by chlorophyll and far-red light (wave length 725–735 nm) is reflected by the leaves. Therefore, a low ratio of red/far-red represents shading which is to be avoided. This situation stimulates a plant to elongate its stem in order to race to the top of the canopy.

5.5 Your Roots, My Roots

To the casual eye, roots may look like a tuft of hopelessly entangled wires, but in fact roots express individuality and have a keen sense of "self" versus "non-self" discrimination. The most convincing way to demonstrate this is to grow two plants in a "fence sitting" situation, in which the roots of each plant straddle two adjacent pots (Fig. 5.2). When compared to a setup where each plant is grown in its own pot, the sharing plants end up with a larger root mass and a lower seed mass, indicating increased root growth at the expense of reproduction, a sign of decreased biological fitness. It looks as if each plant is trying to "elbow" away the other

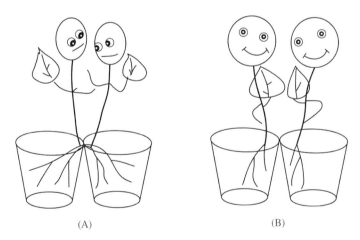

Fig. 5.2. Fence-sitting experiment in beans (*Phaseolus varigaris*) showing effects of habitat sharing by two plants. (A) Two plants each with roots equally divided in two pots (sharing plants). (B) Two plants each with roots in its own pot (owner plants). The amount of soil and its nutrition content are the same in all the pots. Sixty days later, fence-sitters show 150% more root mass than those grown in own pots. However, owner plants show 90% more pod mass, 53% more pods, and 18% more mass per pod, than fence-sitters. The data show that competition in pot-sharing results in poor health for the individuals. [See Note 7; permission Springer-Verlag Dordrecht.]

in order to monopolize the territory and the soil nutrients, even to the detriment of both.[7]

A case of kin recognition has been reported for the Great Lakes beech weed (*Cakile edentula*), in which allocation to roots increases (a sign of competition) when groups of "strangers" (from different maternal families) shared a common pot, but not when groups of "siblings" (kin) shared a pot. As a rule, competition is less intense when the shared individuals are kin. ("Siblings" are plants grown from seeds coming from the same mother plant; "strangers" are plants grown from seeds coming from different plants of the same species.)[8] Subsequent reports confirmed kin recognition in other plant species.[9]

That plants can communicate through roots was elegantly demonstrated in the following experiment (Fig. 5.3). A pea plant (plant A) is grown in such a way that half of its roots are in one pot while the other half are in an adjacent one. The first pot has dry soil whereas the second has normal moisture. The second pot is shared with another plant (plant B), which straddles over another pot that is also shared by plant C. All the pots except the first have normal moisture. Under this arrangement plant A is drought stressed while plants B and C are not. Plant A responds by closure of its stomata to conserve water, a normal reaction

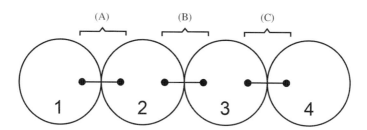

Fig. 5.3. Diagram showing the arrangement of pots to demonstrate relay of root stress cues. Four adjacent pots are positioned in series (from 1 to 4). Pot No. 1 contains soil that simulates drought conditions; all other pots contain soil with normal moisture. A, B and C are plants with split roots, each plant straddling two pots. Response to drought is scored by measuring the closure of stomata (tiny holes on the leaves). All three plants show positive response even though B and C have no direct contact with the dried soil. [See Note 10, Falik; permission Elsevier Science.]

to drought. However, plant B, which is not drought-stressed, also closes its stomata, suggesting that the stress cues put out from the roots are carried from one plant to the next (airborne transmission is excluded in this setup). What is of interest is that, after some delay, plant C also responds to the drought signal, providing evidence that root-mediated messages, like the airborne counterparts, can be serially passed on.[10]

Root exudation is one of the mechanisms of root-root communication. Exudates act within a few millimeters of soil space called "rhizosphere". The rhizosphere represents a highly dynamic front for interaction not only between plants but also between plants and soil microbes and parasites. Root exudation constitutes a significant portion of photosynthetic product and is of considerable carbon cost to the plant. The exudates cover a diverse array of chemicals ranging from small molecules (such as amino acids, organic acids, sugars, phenolics, and other primary and secondary metabolites) to high molecular weight compounds (polysaccharides and proteins). Whether an exudate is beneficial or harmful to the recipient depends on how the message is interpreted. Some exudates are growth facilitators, supporting the growth of other plants. Others induce defense response, either by reducing the susceptibility to pathogens, or by causing the release of volatiles that attract predators of plant-eating insects. (In this sense a parallelism exists between the leaf-warning system above ground and the root-warning system below ground.)[11] When an exudate is toxic to the neighbors, the phenomenon is called allelopathy. The toxic substance mediates competition among plant species. For example, conifers release a mono-terpenoid called alpha-pinene, which inhibits cell proliferation in plants of other species.[12] Other phytotoxins are anti-microbial and insecticides. Ironically, some plants exhibit auto-inhibition by secreting toxins that delay seed germination and growth of seedlings of the same species. It is believed that auto-inhibition is a way for population control in densely populated areas.[11] In addition, root exudates also mediate kin recognition among plants.[13]

Roots evolve mutualistic relationships with certain species of bacteria (rhizobacteria) and fungi (mycorrhizae). Nitrogen-fixing bacteria colonize the roots of some plants (notably the legumes such as soybeans), forming root nodules where the bacteria provide the plant with nitrogen "captured" from the air while the roots serves as a carbon source for the bacteria. Beneficial fungi that live symbiotically with the roots help transport water and minerals (notably phosphorus) to the roots, while taking in carbohydrates from the plants. About 80% of land plants establish mycorrhizal relationship with fungi. The fungal mycelia form a network around a tree and increase the surface absorbing area of roots by 100 to 1,000 times (Fig. 5.4). When roots from two or more plants coexist with the same fungal mycelia, a symbiotic network is formed, which greatly expands the underground communication among plants. This network serves as a conduit not only for nutrition but also for transmission of stress signals. A large underground fungal network increases leaf nitrogen content and reduces the incidence of pathogen-induced root lesions; this beneficial effect is more prominent if the plants are

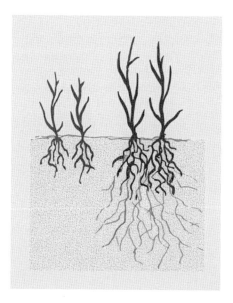

Fig. 5.4. Colonization of plant roots by symbiotic mycorrhizal fungi. The left two plants are grown without the fungi; the right two plants are grown in the presence of the fungi, depicting a continuous network formed by the roots and the fungal mycelia.

closely related.[14] Transfer of stress signal through the root system has been demonstrated in an experiment in which an aphid-infested plant was able to confer resistance to a neighboring tree when the only connection between the two was the fungal network.[15]

A case of nutrition transfer among con-specific plants has been reported for the Douglas fir of the Pacific Northwest. Because of the exceptional height of these trees, younger trees growing in the shade of the older ones have little chance of catching sunlight for photosynthesis. To mitigate this situation, the older trees are able to subsidize their young with nutrients they make. This "mothering" phenomenon was studied with the use of radiocarbon tracer, and the mechanism is believed to involve the underground fungal network.[16]

5.6 How Plants Respond to Pathogen Infection

Plant pathogens include viruses, bacteria, fungi, and nematodes. The following happens rapidly at the site of infection: (a) rapid release of strong oxidative agents to kill the microbes; (b) local accumulation of antimicrobial chemicals; (c) self-induced cell death surrounding the infection site to limit the spread of the disease.

Phytoalexins are small organic molecules that can damage the microbial cell wall, disturb their metabolism, delay their growth, or interfere with their reproduction. Examples of phytoalexins are *flavonoids, resveratrol and alpha-pinene*. Flavonoids are a diverse group (over 4,000 of them) of plant chemicals called polyphenols that, in the normal state, impart characteristic color, odor, and taste to the leaves, flowers, and fruits. Their peculiar taste deters some potential herbivores. They give pigments to flowers to attract insect pollinators. They also serve as light filters to selectively permit useful light to pass through for photosynthesis. During dry seasons, they impart drought resistance to plants. After injury by pathogens or pests, flavonoids accumulate in the lesion margins of leaves to contain the damage.[17] (Humans consume flavonoids for their nutrition value as antioxidants — protection against harmful free radicals.)

Plant stress hormones are a group of small organic molecules produced rapidly in response to injury or infection. *Abscisic acid*, the best known of these hormones, promotes tolerance to stress and has been compared to adrenaline in animals. It is synthesized in all plant parts in response to biotic and abiotic stress and can be translocated from one area to another. Other stress hormones include *salicylic acid, jasmonate, and ethylene*. Together they promote resistance to microbial and insect invasion. They amplify the process of gene activation in times of emergency, priming the entire plant to heighten its defense function. Among the stress hormones, methyl salicylate, methyl jasmonate, and ethylene are volatile, so they can travel airborne from plant to plant, serving as a warning signal. Ethylene is unique in that it also has the normal function of promoting fruit ripening.

"Pathogenesis-related proteins" (PR-proteins) are a group of large molecules having broad antimicrobial functions. They are produced systematically to prime the entire plant for defense when only one area is under attack.[18] *Osmotin* is a prominent member of the PR-proteins. It helps plants resist high osmotic pressure, and is also a potent antifungal agent.[19] The osmotin gene can be activated by a wide variety of factors, including desiccation, NaCl, wounding, ethylene, abscisic acid, viruses, fungi, and UV light.[20]

A sophisticated way for plants to fight viral infection is through *RNA silencing*, also called *RNA interference*. Like animals, plants can cleave viral double-stranded RNA into small pieces of 21-26 nucleotides long, called "small interfering RNA," which are subsequently used to target and destroy any virally produced RNA of the same species. Spreading of the silencing signal from cell to cell and through the vasculature imparts resistance to the whole plant against invasion by the same virus. This type of antiviral strategy is in common with animals (see next chapter).

5.7 Do Plants have Immunity?

Immunity is the defense of an organism against foreign invasion, based on molecular recognition of *self* from *non-self*. Animals have two kinds of

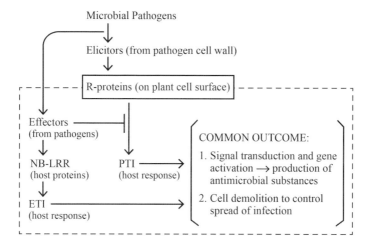

Fig. 5.5. Schematic of plant innate immunity. The dotted-line rectangle outlines the boundary of a plant cell. Microbial pathogens possess special molecules (elicitors) on their cell wall that can be recognized by the surface proteins of a plant cell (R-proteins). The interaction elicits a host response called PTI, which leads to the production of antimicrobial substances and defensive cell destruction. To overcome this, microbial pathogens produce a kind of protein called "effectors" to block the PTI response. To override the block, plant cells also produce a type of protein called NB-LRR. If the NB-LRR matches the "effector" protein, the ETI response is elicited, which bypasses the PTI pathway. The two pathways lead to a common defense outcome. PTI, PAMP-triggered immunity; ETI, effector-triggered immunity; NB-LRR, nucleotide-binding leucine-rich repeat. [See Note 21.]

immunity: (1) the innate type, which is genetically prescribed as a result of evolutionary history; (2) the acquired type, which is determined by the individual exposure in the lifetime of an organism. Plants only have innate immunity, but it is more robust and better developed than that in animals.

Figure 5.5 depicts the two strategies of plant innate immunity. First, plants possess cell surface proteins called R-proteins ("R" for resistance) that can recognize molecules on invading microbial cell wall, called "elicitors." The interaction results in a plant response called "PTI," leading to signal transduction and gene activation, and ending with the production of antimicrobial substances. Simultaneously, some infected cells are encouraged to undergo cell death in order to contain the site of infection. Over the course of evolution, many pathogens developed special molecules called "effectors" in order to sabotage the defense within the plant cell. To counter this attack, plants have produced, over evolution, special types of proteins called NB-LRR that are able to recognize the "effectors" and mount a second defense response called "ETI," leading to the same defense outcome as seen in the original pathway.[21,22] Over many generations, this evolutionary "arms race" between the invaders and the hosts resulted in a great variety of highly specific "effectors" and NB-LRR proteins.

5.8 Rejection of *Self* during Mating

Many flowering plants prevent inbreeding by developing a strategy to distinguish *self* from non-*self* between the male and female sex organs, when both are present in the same plant (hermaphrodites). The recognition is at the molecular level (protein-protein interaction) and involves an intricate self-incompatibility mechanism, wherein pollens from the same plant are recognized and rejected by the pistil (female part of a flower), controlled by determinants encoded in the genes. Self-fertilization is thereby avoided.[23]

Notes and References

1. These secondary plant metabolites frequently have industrial and pharmaceutical uses and are commonly known as "natural products."
2. Weng J-K, Phillipe RN, Noel JP. (2012) The rise of chemodiversity in plants. *Science* **336**: 1667–1670.
3. Karban R, Baldwin IT, Baxter KJ, *et al.* (2000) Induced resistance in wild tobacco plants following clipping of neighboring sagebrush. *Oecologia* **125**: 66–71.
4. Cost C, Heil M. (2006) Herbivore-induced plant volatiles induce an indirect defence in neighboring plants. *J. of Ecology* **94**: 619–628; Heil M, Bueno JCS, (2007) Within plant signaling by volatiles leads to induction and priming of an indirect plant defense in nature., *Proc Natl Acad Sci USA* **104**: 5467–5472.
5. Karban R, Shiojiri K, Ishizaki S, *et al.* (2013) Kin recognition affects plant communication and defense. *Proc R Soc B* **280**: 20123062.

6. BaldwinI T, Halitschke R, Paschold A, *et al.* (2006) Volatile signaling in plant-plant interactions: "Talking Trees" in the Genomic Era. *Science* **311**: 812–815.
7. Maina GG, Brown JS, Gersani M. (2002) Intra-plant versus inter-plant root competition in beans. *Plant Ecology* **160**: 235–247.
8. Dudley SA, File AL. (2007) Kin recognition in an annual plant. *Biology Lett* **3**: 435–438.
9. Murphy GP, Dudley SA. (2009) Kin recognition: Competition and cooperation in *Impatiens*. *Am J Botany* **96**: 1990–1996.
10. Omer F, Mordoch Y, Quansah L, *et al.* (2011) Rumor has it… relay communication of stress cues in plants. *PLoS ONE* **6**: e23625; Falik O, Mordoch Y, Ben-Natan D, *et al.* (2012) Plant responsiveness to root-root communication of stress cues. *Ann of Botany* **110**: 271–280.
11. Bais HP, Weir TL, Perry LG, *et al.* (2006) The role of root exudates in rhizosphere interactions with plants and other organisms. *Ann Rev Plant Biol* **57**: 233–266.
12. Singh HP, Batish DR, Kaur S, *et al.* (2006) Alpha-Pinene inhibits growth and induces oxidative stress in roots. *Ann of Botany* **98**: 1261–1269.
13. Biedrzycki ML, Jilany TA, Dudley SA, Bais HP. (2010) Root exudates mediate kin recognition in plants. *Communicative and Integrative Biol* **3**: 1–7.
14. File AL, Klironomos J, Maherali H, Dudley SA. (2012) Plant kin recognition enhances abundance of symbiotic microbial partner. PLoS ONE **7**: e45648.
15. Babikova Z, Gilbert L, Bruce TJA, *et al.* (2013) Underground signals carried through common mycelial networks warn neighbouring plants to aphid attack. *Ecology Letters* **16**: 835–843.
16. Simard SW, Perry DA, Jones MD, *et al.* (1997) Net transfer of carbon between ectomycorrhizal tree species in the field. *Nature* **388**: 579–582; Simard SW. (2012) Mycorrhizal networks and seedling establishment in Douglas-fir forests. In Southworth D. ed. *Biocomplexity of Plant-Fungal Interactions.* John Wiley, Chap. 4, p. 85–107.
17. Treutter D. (2006) Significance of flavonoids in plant resistance: A review. *Environmental Chem Lett* **4**: 147–157.
18. Kitajima S, Sato F.(1999) Plant pathogenesis-related proteins: Molecular mechanisms of gene expression and protein function.*J Biochem* **125**: 1–8.

19. Singh NK, Bracker CA, Hasegawa PM, *et al.* (1987) Characterization of osmotin: A thaumatin-like protein associated with osmotic adaptation in plant cells. *Plant Physiology* **85:** 529–536; Selitrennikoff CP (2001) Antifungal proteins.*Applied and Environmental Microbiol* **67:** 2883–2894.
20. LaRosa PC, Chen Z, Nelson DE, *et al.* (1992) Osmotin gene expression is post-transcriptionally regulated. *Plant Physiol* **100:** 409–415.
21. Dodds PN, Rathjen JP. (2010) Plant immunity: Towards an integrated view of plant-pathogen interactions. *Nature Rev Genetics* **11:** 539–548.
22. Plant immune recognition proteins have a remarkable sequence analogy to their mammalian counterparts, a testament to evolutionary commonality in *self*-recognition.
23. Kubo K, Entani T, Takara A, *et al.* (2010) Collaborative non-self recognition system in S-RNase–based self-incompatibility. *Science* **330:** 796–799.

Chapter 6
The Animal *Self*: Molecular Recognition

Molecular markings delineate bodily self.

Overview: *In a most subtle and intricate manner, the animal body zealously guards itself against intrusion by other "selves," using a mechanism known as immunity. Different aspects of immunity define self in various degrees of stringency. In innate immunity, self is defined categorically in a phylogenetic sense — for example, animals against microbes. In adaptive (acquired) immunity, the distinction is further refined so that antibodies against one species of pathogen are different from those against others — for example, diphtheria versus tuberculosis. However, the most stringent definition of molecular self appears in the recognition of MHC (major histocompatibility complex) protein by T lymphocytes, for here immunity recognizes differences not only among species, but also among individuals of the same species, including siblings. The animal body is defined by its immunological self, which is based on recognition of the detailed structure of body proteins. The rejection of organs when transplanted from one person to another is a prime example of "self" versus "non-self" discrimination.*

6.1 Drama in a Glasgow Hospital

In 1943 a lady was admitted to the Glasgow Royal Infirmary for an extensive burn, involving the right side of her back and part of her right arm. For this she underwent skin transplantation for the burnt area. One part was replaced with her own skin taken from a normal site, and another was transplanted with skin from her brother. Both skin grafts survived initially. But, surprisingly, by the fifteenth day, whereas the transplant derived from her own body (autograft) continued to thrive, that from her brother (allograft) began to degenerate and eventually sloughed off. Here lies the puzzle, why would the body tolerate a piece of its own skin but reject one from someone else?[1]

Gibson and Medawar, who observed this landmark phenomenon, reasoned that the host refused to accept the "foreign" skin because it was "non-*self*". For the first time science realized that the human body, and any animal body for that matter, is endowed with an ability to recognize *self*. A transplanted tissue or organ is non-*self* and therefore has to be destroyed, just as an invading bacterium should be removed by the immune system. From then on the concepts of *self* and *immunity* became inseparable in modern medicine. Since immunity is founded on molecular pattern recognition, there must be a strong molecular basis for the body to tell *self* from non-*self*.

This story reminds us of an earlier example of immunity: blood transfusion reaction. It was known for some time that if blood is taken from one species and transfused to another, the red blood corpuscles from the donor are clumped and broken up, leading to serious illness and death of the recipient. The same problem frequently happens when blood is transfused between humans. In 1901, Landsteiner discovered that human blood can be classified into four groups, called A, B, AB and O.[2] Transfusion within the same group gives no problem, but transfusion between groups leads to variable outcomes: group A can receive blood from O, but not from B or AB; group B can receive from O but not A or AB; group AB can receive from A, B or O; group O cannot receive blood from any group other than its own type (Table 6.1 and Fig. 6.1). This complex relationship can be explained in molecular terms. Each blood type has a unique molecular make-up on the surface of its red blood cells. The A, B or O types are determined by surface polysaccharides. Each blood group perceives the others as foreign and attempts to destroy the transfused red cells with host antibodies. No antibodies are produced against the markers if the host already possesses the same markers. Thus no reaction takes place when transfusion is within the

Table 6.1. Major Human Blood Types

Blood type name	Group A	Group B	Group AB	Group O
Antigens in red cell	A	B	A&B	None
Antibodies in serum	Anti B	Anti A	None	Anti A&B

Note: In transfusion, blood types are named after the antigens in red blood cell. A person's blood serum contains antibodies against antigens that his own red cells do not have, but not against antigens that his own red cells have. Transfusion reaction occurs when the recipient's serum antibodies react against the red cells from an incompatible donor.

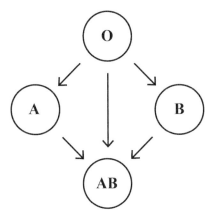

Fig. 6.1. Blood type compatibility diagram. The arrows show how blood can be donated from one person to another with different blood types. In blood transfusion, it is the red blood cells from the donor that is of concern. Incompatibility results from the reaction of the antibodies in the recipient's serum toward the donor's red cells. [See Table 6.1 for further explanation.]

same blood group. In the business of transfusion, persons belonging to the same blood type behave as if they were a collective *self*.[3]

Another example of immunity is the defense against pathogens. The world we live in is a very dangerous place, for we are threatened not only by predators larger than us, but also by invisible microbial pathogens lurking around, waiting for a chance to strike and consume us. Without our knowing it, every cell in our body is constantly vigilant of these tiny intruders. Historically, vaccination as a way to prevent infection was practiced in the East as early as the tenth century.[4] It was brought to modern medicine in 1796 when Edward Jenner started to inoculate cowpox for protection against smallpox. The concept came from the common observation that milkmaids who had contracted cowpox did not get smallpox. Immunology received a boost with the subsequent discovery of microorganisms and their connection to disease; and with the discovery of special proteins called antibodies circulating in the blood that are capable of neutralizing these pathogens.

Today, we realize that all of immunity, with its ramifications in transplantation, transfusion, vaccination, and defense against pathogens, boils down to the principle of "self" versus "non-self" discrimination on the molecular level.

6.2 Innate Immunity is Evolutionarily Very Old

There are two types of animal immunity: the inborn or *innate immunity* which is rapid in action but lacks specificity, and the acquired or *adaptive immunity* which is slow but highly specific. Both are defense against invaders. The distinction is that innate immunity is shaped by evolution and passes down from generation to generation through the genes, whereas adaptive immunity is developed (acquired) only during the lifetime of an animal.[5]

Innate immunity appeared very early in evolution, being present in all multicellular organisms, including vertebrates and invertebrates, and even plants (see *Chapter* 5). Innate immunity depends on recognition of common features in pathogens, the so-called "danger signals." The recognition is genetically programmed and does not need any prior encounter with the invaders. Its function is to provide a quick and immediate halt of the invading microbes. Without innate immunity, a single pathogenic bacterium in blood may double every hour and end up with 20 million progeny in a day, causing a full-blown infection. Innate immunity keeps the infection in check while awaiting the more powerful adaptive immunity to kick in, which takes about a week's time.

Innate immunity in animals consists of three lines of defense. *First*, there are physical barriers in the form of skin, epithelial lining, and cell membrane (cell wall in plants). These barriers are frequently fortified by chemical agents that discourage pathogen growth, such as acidity in the stomach and lysozyme in tears. Many epithelial cells protect themselves with a mucus layer that contains antibacterial agents, the most abundant among which are the *defensins*, polypeptides of about 12-50 amino acids long. Defensins are produced not only by epithelial cells but also by neutrophilic leukocytes and are partly responsible for their bactericidal mechanism. Defensins belong to a large group of innate host defense peptides (also known as *antimicrobial peptides*) found in all animals and plants. They exhibit a broad-spectral activity that kills bacteria, fungi, protozoa, nematodes and some viruses.[6]

The *second* line of defense is the ability of all cells, when infected by a microbe, to attempt to degrade the invader with a variety of digestive enzymes inside the cell. The *third* line of defense consists of special proteins and phagocytic cells, dedicated to actively attacking the invaders based on molecular recognition. Molecular features common to pathogens are perceived by animal cells as foreign.[7] They trigger an array of innate immune responses after binding to host receptors, collectively called *pattern recognition receptors*. These receptors are found in many places: in the blood, on the cell surface, and inside the cell. The soluble receptors in the blood are part of the "complement system," a cascade of protein chain-reactions involved in direct and indirect destruction of the pathogens. The cell-associated receptors stimulate phagocytosis (ingestion of foreign bodies) and also activate the gene to initiate a series of immune responses. Both animals and plants possess these receptors, the best known of which are the *Toll-like receptors*.

6.3 Cells Involved in Innate Immunity

The major cellular players in innate immunity are (see Table 6.2): macrophages, neutrophils, dendritic cells and natural killer cells (NK cells).

Table 6.2. Cells Participating in Animal Immunity

A. Cells involved in innate immunity:
These cells mount an immediate, general response to invaders without reference to specific antigens.
1. **Macrophages** – Large phagocytic cells present in all tissues. Main function is "garbage disposal" – engulfing and digesting foreign bodies and cellular debris. Recruit neutrophils to the site of infection.
2. **Neutrophils** – Major phagocytes in the blood. "Foot soldiers" against invading bacteria. Arrive rapidly at the scene of infection and destroy pathogens on the spot.
3. **Dendritic cells** – Main function is to present foreign antigens to the T lymphocytes to trigger adaptive immunity. Serve as immune "sentinels" by patrolling the whole body in search of pathogens. Digest pathogen proteins and display the peptide fragments on cell surface for presentation to T cells.
4. **Natural killer cells** (NK cells) – Innate immune cells circulating in the blood. Eliminate virus-infected cells and certain tumor cells.

B. Cells involved in adaptive (acquired) immunity:
These cells mount a delayed response directed to specific foreign antigens following provocation.
1. **B lymphocytes** –
Main function is to produce antibodies. Activated upon exposure to circulating foreign antigens. Fully mature antibody-secreting B cells are called plasma cells.
2. **T lymphocytes** –
   a. Cytotoxic T cells: Activated by dendritic cells. Seek out virus-infected cells and destroy them, thus the nickname "killer T cells."
   b. Helper T cells: Activated by dendritic cells. Main function is to enhance immune response by stimulating B cells and macrophages.
   c. Regulatory T cells: Activated by dendritic cells. Main function is to suppress immune response when immunity is no longer needed.
3. **Memory cells** –
These are activated B and T cells that do not participate in attacking the invading pathogens but are reserved for future activity when invaded again by the same pathogens (re-infection). Immune response to re-infection is much faster and robust than first infection.

Note: Immune cells circulating in the blood make up most of the white blood cells (leukocytes).

Except for NK cells, they recognize pathogens through their *pattern recognition receptors*, though each plays a different role in fighting the invaders.

Macrophages and neutrophils are both phagocytes — cells capable of engulfing pathogens. Mature macrophages patrol all tissues in the body and function as universal "garbage collectors," removing all unwanted materials on the way (their large size seems to make them suitable for this task). Upon encountering pathogens, macrophages also send out signals to recruit other immune cells such as neutrophils to the scene.

Neutrophils are combat soldiers at the frontline. Present in overwhelming number in the blood, they are quickly attracted to the infection site and engage in frenzied killing. They take up the pathogens, digest them with enzymes, and literally "burn" them to death with strong oxidizing agents. Their job is to destroy the invaders as quickly as possible, frequently at the expense of their own life. (Pus cells found in infected wounds are dead neutrophils after doing their job of defense.)

Dendritic cells are so-called because of their star-shaped, branched appearance (not to be confused with the dendrites of a neuron). Rather than directly attacking the pathogens, they gather chemical information about the antigen and present it to the adaptive immune cells (lymphocytes) for a delayed but powerful response. This is accomplished by digesting the foreign proteins and displaying the resulting peptides on their cell surface. Considered the most effective *antigen-presenting cells*, the dendritic cells through their "scouting" function serve as a link between the innate and the adaptive immune response.

Natural killer cells (NK cells), the fourth type of cells in innate immunity, do not recognize or attack the invading pathogens. Instead, they can directly detect host cells that are infected with virus. They then destroy the virus-infected cells in order to prevent the spread of infection. They also destroy certain tumor cells and are thought to be involved in immune surveillance against malignancy.

6.4 Special Strategies Against Viruses

Since viral particles are little more than pockets of nucleic acids, they lack proteins and other markers that can easily be distinguished by host cells as non-*self*. To remedy this, the host cells develop special strategies by focusing on viral nucleic acids. One way to single out viral DNA is the abundance of the unmethylated "CpG" motif, a property shared by both viruses and bacteria, which is recognized by host cell's Toll-like receptors. Another way to deal with viruses is through a process called RNA interference. In this process, mammalian cells take advantage of the fact that viral RNA inside the infected cells goes through a double-stranded stage, whereas mammalian RNA are always single-stranded. Thus, double-stranded viral RNAs are recognized as foreign by host cells and are degraded into small pieces. These viral RNA fragments then bind to the majority of intact viral RNA (which are single-stranded) if their sequences match. The binding triggers a destructive mechanism to get rid of the virus. Furthermore, viral RNA fragments stimulate the infected cells to secrete *interferons*, powerful signals that alert many cells in the body to mount an all-out antiviral response, through induction of more than 300 different genes. (Note that the anti-viral mechanism through RNA interference is also found in plants.) Animal cells also recognize foreign DNA (usually from a virus) in the cytoplasm, through a sensor called cGAS (cyclic GMP–AMP synthase). cGAS is an enzyme that produces the dinucleotide cyclic GMP–AMP, which sends a message to the nucleus to activate the production and secretion of interferons, leading to an all-out anti-viral response, as described above for viral RNA.[8]

6.5 Adaptive (Acquired) Immunity Came in Late but Robust

Adaptive immunity appeared in evolution less than 500 million years ago, and is present only in vertebrates. Adaptive immunity deals with antigens encountered during the lifetime of an animal, needs a lag period of one to two weeks to act, is highly specific in terms of molecular recognition, and has a long-lasting effect. Each repeated exposure to the same antigen strengthens the immune response, suggesting the presence of immune "memory."

The major players of adaptive immunity are the lymphocytes (see Table 6.2), which are of two types: the B cells which exclusively make antibodies and release them into the blood stream, creating a long-distance systemic effect (thus the term "antibody-mediated immunity" or "humoral immunity"); the T cells which act by cell contact (cell-mediated immunity) for the destruction of the infected cells. T cells also secret cytokines for remote regulation of immune function.[9]

There are three types of T cells: the cytotoxic T (also called "killer T"), the helper T, and the regulatory T. All T cells are activated by exposure to the antigen presented to them by antigen-presenting cells, in particular the dendritic cells. *Cytotoxic* T destroy virus-infected host cells. *Helper* T do not kill cells but instead secrete cytokines to enhance B cell and macrophage functions. *Regulatory* T counteracts helper T by shutting down immune response once infection is over.

6.6 Gene Scrambling Leads to Antibody Diversity and Specificity

Antibodies are exclusively produced by B lymphocytes. They ward off infection in the following manners: (1) binding of antibodies to extracellular pathogens can inactivate the invaders; (2) binding of antibodies to extracellular toxins can neutralize their toxicity; (3) antibodies mark the invading bacteria for destruction by *complements*, a family of special immune proteins; (4) antibodies mark the invading bacteria for phagocytosis by macrophages and neutrophils.

For antibodies to be useful, they must be produced in great varieties, yet each type has to be highly specific for binding only one kind of antigen. How the lymphocytes achieve these two seemingly opposing goals is a great wonder of nature. Before explaining how antibodies are made, it will be helpful to clarify the concept of antibody-antigen interaction. Antibodies are a special type of protein capable of binding other molecules, referred to as antigens, usually another macromolecule like protein or polysaccharide (such as those on a bacterial surface). The part of an antibody involved in binding is called the "antigen binding site," and the part of the antigen that binds to the antibody is called "antigenic determinant" or "epitope." The exact fit of the two sites in a "lock and

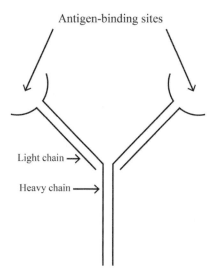

Fig. 6.2. Schematic view of a typical antibody, IgG (immunoglobulin G). Antibodies are proteins made up of amino acid chains. IgG comprises a pair of heavy chains and a pair of light chains. Both chains contribute to the antigen-binding sites.

key" manner dictates how strong and specific the binding is. The stronger the binding, the more effective is the antibody. Antibody-antigen binding takes place within a stretch of ten amino acids in a protein molecule; sometimes an antibody can tell the difference between two antigens that differ in only one amino acid. Figure 6.2 shows schematically the structure of an antibody.

Common sense leads to the thinking that antibodies derive their specificity following contact with the antigens, but this is not quite the case. The surprising fact is that a repertoire of antibodies, covering an enormously wide range of specificity, is ready-made by the naïve B cells before they are even exposed to any foreign antigen. This is achieved by a unique way of random gene recombination occurring in the nucleus of the lymphocytes. By one estimate, an animal can generate a potential repertoire of 10^{12} specific antibodies, enough to bind more types of antigens than it can ever expect to encounter in a lifetime. On the other hand, each lymphocyte is uni-potent, i.e., it is dedicated to producing

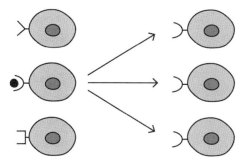

Fig. 6.3. Clonal selection of B lymphocytes. An immature B cell whose preformed antibody on the cell surface matches a foreign antigen (dark dot on left) starts to proliferate and secret antibody into the blood stream.

only one type of antibody. Thus, within the entire population of naïve lymphocytes, only a tiny fraction (perhaps even one cell) actually carries antibodies directed against a particular antigen. Nevertheless, upon encountering a matching antigen, this single cell starts to proliferate to form a clone that produces the specific antibody in large quantities. This phenomenon is called "clonal selection."

Figure 6.3 shows how clonal selection works. Note that the naïve B lymphocytes display their preformed antibodies as "receptors" on the cell surface. Once activated by a foreign antigen, the single expanded clone secretes the specific antibody whose structure corresponds to the activated surface receptor.

6.7 MHC protein: A Personalized Molecular Signature

Like B cells, naïve T lymphocytes express surface "receptors" that bind antigens, but unlike B cells, which release their receptors to become antibodies, T cell receptors stay on the cell surface and are never secreted into the blood stream. T cell receptors are as diverse and specific as those of B cells and are also a result of gene shuffling and clonal selection. Whereas B cells can recognize antigens directly, T cells recognize only those antigens that have been processed by "professional" antigen-presenting cells (such as dendritic cells) and displayed on the latter's cell surface. In other words, cell-to-cell contact is needed for T lymphocyte function.

MHC (acronym for "major histocompatibility complex") is a cluster of glyco-protein molecules located on the surface of all mammalian cells. When a cell is invaded by a virus, viral proteins are digested by the host cell into peptides (protein fragments) and displayed on the surface as a complex with MHC, called peptide-MHC complex. Dendritic cells present this peptide-MHC to cytotoxic T lymphocytes and turn them into killer cells. When an activated cytotoxic T subsequently encounters a virus-infected cell (a non-immune cell) whose surface displays the same peptide-MHC, the infected cell is destroyed (Fig. 6.4).

Three features of MHC are crucial to the understanding of molecular *self*: (1) the same MHC is present on the surface of all cells of an individual whether or not they are infected (i.e., it is a normal cell

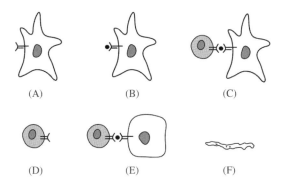

Fig. 6.4. Activation and function of cytotoxic T lymphocytes. (A) A dendritic cell (star shaped) with MHC protein on the surface. (B) A dendritic cell displaying a piece of viral peptide (shown as a dot) on the cell surface as peptide-MHC complex. (C) A dendritic cell presenting a viral peptide to the receptor of a cytotoxic T lymphocyte (round cell). (D) An activated cytotoxic T cell (now a killer cell) circulating in the blood. (E) A killer T cell encountering a virus-infected non-immune cell (square shaped) whose surface peptide-MHC complex matches that of the dendritic cell. (F) Demise of the virus-infected cell. Note that all cells (immune and non-immune) that are infected by virus display the viral peptide as the peptide-MHC complex, but only the immune cells (shown here as a dendritic cell) are capable of triggering T lymphocytes to kill.

component); (2) for reasons beyond the scope of this book, *each individual's MHC is unique (in amino acid sequence) and is not shared by any other person*, even among siblings, much like every person has his own set of finger prints; (3) the T cells of a person will mount an attack on any cell that carries a non-*self* MHC. In this sense, MHC serves as a stringent molecular marker for the physical *self*.[10]

Recall the story in the beginning of this chapter, in which a transplant recipient rejected her brother's skin graft but kept her own, leading Medawar to suspect that organ rejection is a form of immune attack. For this historical reason, the term "major histocompatibility complex" (MHC) was coined with reference to tissue transplantation ("histo" is the Greek equivalent of "tissue"), before it was later realized to be also involved in anti-infection.

The truth is, pathogens and foreign tissues have one thing in common: they both are molecular non-*self* intruders and should be ferociously rejected by the host. Individual variation in MHC is exceedingly intricate, and it is extremely rare for two genetically unrelated persons to have the same MHC protein. In practice, only identical twins make a perfect match in transplantation. All other matches, including siblings, show various degrees of incompatibility, requiring immunosuppressive agents to mitigate rejection. Table 6.3 shows rejection of a skin graft when the donor comes from a person (or animal) of different genetic makeup.

Table 6.3. Outcomes of Skin Transplantation

Donor	Recipient	Outcome
Syngeneic	First time	Acceptance
Allogeneic	First time	Rejection
Allogeneic	Second time	Accelerated rejection

This table shows tolerance of *self* (syngeneic donor) and rejection of non-*self* (allogeneic donor) in skin grafting. "Syngeneic" refers to two individuals with identical genetic makeup; "allogeneic" refers to those belonging to the same species but different in genetic makeup. The third row shows that repeated transplantation of foreign skin from the same donor leads to accelerated rejection of the graft.

6.8 When *Self* Attacks *Self*: Autoimmunity

How does the immune system avoid attacking its own cells? While working hard to eliminate non-*self* intruders, the immune system of an animal must walk the fine line of not attacking itself. This important task is accomplished by a process called "negative selection" during the developmental stage of the immune system. By a mechanism not totally understood, the bone marrow (where B cells mature) and thymus (where T cells mature) manage to express an entire spectrum of *self*-proteins belonging to the animal. Young lymphocytes with "receptors" (surface antibodies) that bind strongly to these *self*-proteins are eliminated (negative selection). Only lymphocytes that survive this initial pruning are released to the lymph nodes, where they will be activated by non-*self* antigens through positive selection. In effect, the bone marrow and thymus serve as "training camps" where lymphocytes are "educated" to distinguish *self* from non-*self*.

Like other processes in nature, this complicated system of cell selection is not perfect, and sometimes aberrant outcomes do occur, leading to either overreaction (autoimmune diseases) or under-reaction (immunodeficiency). Examples of autoimmune diseases include myasthenia gravis, in which auto-antibodies attack acetylcholine receptors in neuromuscular junctions, resulting in muscle weakness; rheumatoid arthritis, where T lymphocytes attack joint tissue lining, leading to inflammatory joint disease; and multiple sclerosis, where T cells destroy myelin proteins, causing multiple neurologic symptoms. One way to treat these diseases is to give immunosuppressive drugs, such as corticosteroids, cytokines, and agents toxic to immune cells. Some of the autoimmune diseases are rare, but as medical progress prolongs life span, many rare autoimmune disorders surface and become more prevalent.

6.9 When *Self* Tolerates another *Self*: Pregnancy

One instance when *self* naturally tolerates non-*self* is the mother-fetus relationship. Fetuses inherit some of the paternal proteins, which are foreign to the mother. Normally the conflict is avoided by the non-mixing of maternal and fetal circulations. Nevertheless, some of these foreign cells may escape from the fetal to the maternal circulation (through small tears in the placenta during delivery), and this induces antibodies from the mother against the fetus. During subsequent pregnancies, some of these antibodies may pass through the placental barrier and attack the second fetus, if it also expresses the same offending antigen. A common example of this situation is blood group incompatibility, leading to hemolysis (breakage of red blood cells) and jaundice in the newborn. In severe cases, blood transfusion may be needed to save the baby.

The immune system can be "coaxed" to accept a foreign antigen by introducing the antigen in small doses over a long period of time. This phenomenon, called *anergy*, occurs naturally in the body against some *self* proteins as part of immune system maturation. The same principle can be used to desensitize allergy, a type of immune reaction.

In an experimental setting, an animal can be fooled to accept an organ transplant by introducing the donor cell to the recipient when the latter is at a very young age, during a critical period when the immune system is being "trained" to discriminate between *self* and non-*self* (Fig. 6.5).

Fig. 6.5. How *self* can be "fooled" to accept non-*self*. A skin graft was successfully transplanted from an adult brown mouse to an adult white mouse. The white mouse tolerated the foreign skin because earlier (at the time of birth) it received an injection of bone marrow cells (the precursors of all immune cells) from the brown mouse, thus changing its immunological makeup. [See Note 14; permission Mosby, Inc.]

6.10 Graft-Versus-Host: How the Guest takes over the House

Under certain circumstances, a most dramatic scene plays out when the transplanted tissue (non-*self*) attacks the host (*self*) and wreaks severe havoc on the body. This can happen in bone marrow transplantation. Certain medical conditions require replacement of bone marrow as a treatment. If the donor's bone marrow, which contains mature T lymphocytes, is immunologically incompatible with the host, the guest lymphocytes can mount an attack on virtually every part of the host, leading to dire consequences. In this instance, the donor is the offensive and the recipient becomes the defensive, a reverse of what usually occurs in other types of tissue transplantation.

6.11 When *Selves* Play Hide-and-Seek: Parasite-Host Interaction

In many instances the host and the invading organism engage in a war of immunological hide-and-seek. One example is African trypanosome, the protozoan parasite that causes a fatal disease called "sleeping sickness." The parasites evade their host's immune recognition by constantly changing their surface proteins, a process called antigenic variation. A given trypanosome population can sequentially express hundreds of these variant surface proteins in order to confuse and escape the host's immune recognition. This is an example of changing the molecular identity of *self* by an organism.[11]

6.12 Immune Surveillance: Elimination of the Renegade

The strongest reason to suspect the presence of immune surveillance against malignant tumors is that people who are immunologically compromised (such as those taking immune suppressive drugs) are more likely to develop cancer. When normal cells turn cancerous, chemical alterations occur, among which is a decrease in MHC protein (a marker of *self*) on the cell surface. Natural killer (NK) cells consider these MHC-deficient cells as non-*self* and destroy them.[12] People with deficient NK cell activity have been associated with increased cancer risk, such as prostate cancer and melanoma. Experiments in mice also show that NK cells protect the host against certain chemically induced tumors.[13]

Notes and References

1. Gibson T, Medawar PB. (1943) The fate of skin homografts in man. *J Anatomy* **77**: 299–310.
2. Landsteiner K. (1901) Uber agglutinationserscheinungen normalen menschlichen blutes. *Wiener Klin Wschr* **14**: 1132–1134.
3. Although organ transplantation and blood transfusion both involve immunity and *self*-recognition, there are differences between the two. First, in transplantation the immune reaction is cell mediated whereas in transfusion it is antibody mediated. Second, organ transplantation has a more stringent *self*-recognition than transfusion, as the former recognizes differences from individual to individual, whereas the latter, between groups of individuals. Third, in organ transplants the host encounters the foreign molecules for the first time, and therefore a lag period is needed to elicit a full immune response; in blood transfusion the response is immediate because the antibodies against the antigens are already in the blood, as the recipient has earlier been exposed to the same antigens found in food and common microbes. Fourth, in transfusion of whole blood, only the donor red cells are of importance, as they are the major cellular elements in the blood. Plasma from the donor is greatly diluted and does not play a significant role in the overall transfusion reaction.
4. In China, powdered smallpox scabs were blown up the noses of the healthy. The patients would then develop a mild case of the disease and from then on were immune to it. See: Temple R. (1986) *The Genius of China: 3000 Years of Science, Discoveries and Inventions.* Simon and Schuster, New York, pp. 135–137.
5. Although innate immunity is adaptive in the evolutionary sense, the term "adaptive immunity" is reserved for the acquired type of immunity.
6. Zasloff M. (2002) Antimicrobial peptides of multicellular organisms. *Nature* **415**: 389–395; Giuliani A, Giovanna P, Nicoletto SF. (2007)

Antimicrobial peptides: an overview of a promising class of therapeutics. *Cent Eur J Biol* **2**: 1–33.

7. For example, bacterial proteins are special in containing formylated methionine on one end. Bacterial DNA differs from animal DNA in being rich in unmethylated "CpG motif." Besides, bacteria possess cell walls and flagella, which are absent in animals. Components of bacterial cell walls such as peptidoglycan, lipopolysaccharide (LPS), and teichoic acid, and molecules unique to fungal cell walls such as mannan, glucan and chitin, are recognized by animal cells as non-*self*.

8. Sun L, Wu J, Du F, *et al.* (2013) Cyclic GMP-AMP synthase is a cytosolic DNA sensor that activates the type I interferon pathway. *Science* **339**: 786–791.

9. Both B and T lymphocytes originate in the bone marrow, but T cells, in addition, have to pass through the thymus to reach maturity, hence the name "T".

10. There are two types of MHC: MHC-I and MHC-II. Whereas MHC-II is found only in specialized antigen presenting cells (such as dendritic cells), MHC-I is present on the surface of all cells except red blood cells. MHC-I is the one that is relevant in recognition of *self* during organ or tissue transplantation. Since each individual has a unique MHC-I, transplantation of organs or tissue from one person to another triggers T cell-mediated immune rejection by the host. The individual difference in MHC-I is due to variation in amino acid sequence of the protein, a phenomenon called genetic polymorphism.

11. Donelson JE. (2003) Antigenic variation and the African trypanosome genome. *Acta Tropica* **85**: 391–404.

12. Ljunggren H-G, Kärre K. (1990) In search of the "missing self": MHC molecules and NK cell recognition. *Immunology Today* **11**: 237–244.

13. For general references for this chapter, see the following: Alberts B, *et al.* (2008) *Molecular Biology of the Cell*. 5th ed. Garland Science, New York; Murphy K. (2012) *Janeway's Immunobiology*. 8th ed. Garland Science, New York; Parham P. (2009) *The Immune System*. 3rd ed. Garland Science, New York.

14. Roitt IM. (1988) *Essential Immunology*. 6th ed. Blackwell Mosby, Oxford, UK.

Chapter 7
The Animal *Self*: Neurobehavioral Correlates

If there is a more wonderfully complex structure in the universe, I do not know of it.

— Seymour S. Kety[1]

Overview: *Animals differ from plants in that they are capable of moving around in search of food and mates and to avoid hostile environments. In order to take goal-directed actions, the many cells in the animal body must coordinate in an efficient way. The nervous system, with its complex communication network, makes this possible. The building block of the nervous system is the neuron. Two or more neurons are interconnected through a specialized contact called the synapse, where a very narrow gap separates the two ends. The gap, called a synaptic junction, relies on chemicals for the message to get across, providing a site for intricate regulation and modulation. The prevailing theory is that synaptic plasticity is the basis of learning and memory.*

The nervous system evolved from a diffuse nerve net to a centralized structure called the brain. Primitive brains are concerned with the essential functions of life: survival and procreation. To this, evolution added the emotional component, providing the driving force for action. Topping all these is the cerebral cortex, which regulates the primitive urge. For example, the inhibition of an emotional impulse makes possible deferred gratification, an act that may eventually bring in a greater reward than what can be obtained from immediate satisfaction. The cerebral cortex permits higher functions such as reasoning and abstraction, and leads to emergence of the mind. The overall function of the brain imparts unity to the behavioral self. Disruption of the brain leads to the fragmentation of mind, and self.

■ 127

7.1 Neurons: Building Blocks of the Nervous System

A single-celled organism such as an amoeba moves around in search of food. Such simple locomotion needs internal coordination of the contractile elements, so that when one end extends forward the opposite end retracts. Now, imagine a hypothetical two-cell organism. If the two cells pull in opposite directions, locomotion would be impossible and the animal would starve. The two cells need to communicate and coordinate, probably by ion fluxes across plasma membranes. But visualize a multicellular animal with hundreds and thousands of cells. Clearly, communication by contact one cell at a time would be extremely slow and cumbersome. In this situation evolution came up with specialized, dedicated cells for rapid coordination over some distance. These are the *neurons*, the building blocks of the nervous system. Although throughout evolution the nervous system varies in complexity, from the simple nerve net of a coelenterate to the billions of specifically interconnected neurons of the human brain, individual nerve cells share many common features and their functions are basically the same.

Figures. 7.1 and 7.2 show a typical neuron and its connections. A neuron consists of three major parts: (1) a cell body or *soma* that contains the nucleus and other organelles; (2) numerous short, branching processes called *dendrites* that receive messages from other neurons; (3) a single long *axon* that conducts messages away from the soma. An axon sends out branches at the end of its course that terminate as nerve endings, forming connections (*synapses*) with the dendrites and cell bodies of other neurons. Neuronal messages traverse down the axon in the form of *action potentials*, which are waves of electrical perturbations caused by ionic fluxes across the plasma membrane (inward flow of sodium followed by outward flow of potassium). This is made possible by a unique property of the axonal plasma membrane called "excitability" (also present in muscle cells). Like other plasma membranes, the axonal membrane is a capacitor that separates charges across its two surfaces, forming an electrical potential. But, unlike other membranes, once the axonal membrane potential reaches a certain threshold (from a resting

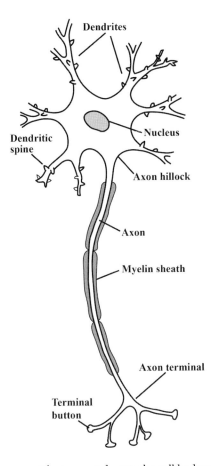

Fig. 7.1. A typical neuron. The top part depicts the cell body with multiple dendrites; the middle part, the axon with myelin sheath; the bottom part, the nerve terminals that synapse with other neurons. Small protrusions from the dendrites are dendritic spines, the sites of contact with incoming axons. The information flow is from the dendrites to the cell body and then to the axon.

potential of −70 millivolts to −50 millivolts, measured from the inside), it automatically triggers a transient, sudden influx of sodium ions, raising the potential to about +30 millivolt (called the action potential) before returning to the resting level (Fig. 7.3). The propagation of the action potential (nerve impulse), once started, is an all-or-none phenomenon, its amplitude and velocity being fixed for a given axon. Its frequency, on the other hand, increases with the strength of stimulation. The speed

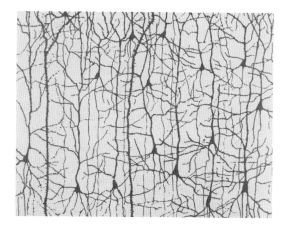

Fig. 7.2. Interconnection of numerous neurons in the human cerebral cortex as revealed by silver stain. [Source: S. Ramon y Cajal, public domain.]

of a nerve impulse ranges from 0.1 to about 100 meters per second, depending on the diameter of the axon (the bigger the faster) and the degree of insulation (the more myelination, the faster). This is a very slow process compared to an electric wire. However, the slowness "buys time" for modulation and subtle adjustments, a feature not possible with a metallic conductor.[2]

The size of the nerve cell body ranges from a few thousandths of a millimeter to a tenth of a millimeter, whereas the length of an axon can be as long as a meter in a person. It has been estimated that the total length of all the axons in a human brain exceeds 100,000 miles. A neuron can receive incoming messages from as many as a thousand neurons, and can transmit to as many others. The complexity of neuronal connections in a single human brain dwarfs the relationship among all the stars in the universe. The synapse, which connects a nerve terminal to a dendrite, consists of a pre-synaptic portion (axonal side, releasing neurotransmitters), a post-synaptic portion (dendritic side, receiving neurotransmitters) and an intervening gap called a synaptic cleft (Fig. 7.4). Usually, the portion of the dendrite that forms a synapse is a specialized area (a small protrusion) called the *dendritic spine* (see Fig. 7.1). Chemical communication between neurons is a one-way street, made possible by the release

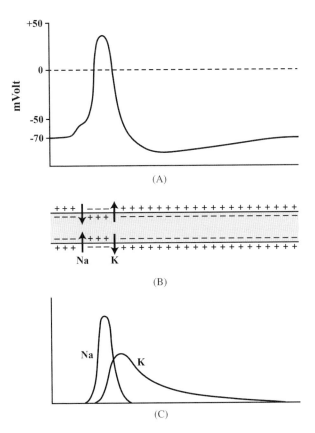

Fig. 7.3. Diagram of an action potential with abscissa representing time course. (A) Voltage change (measured inside the axon) from −70 (resting) to +30 (peak) millivolts, then down to −80 before going back to the resting value. (B) Change in membrane charge from inside negative (resting) to inside positive (result of sodium influx), which returns to the resting stage following potassium efflux. The change from negative to positive is called depolarization; the opposite, hyper-polarization. (C) Changes in sodium and potassium permeability across axonal membrane corresponding to changes in potential. Note that while post-synaptic membrane potential changes take place in the dendrites and cell body, action potential starts only at the axon hillock when local membrane potential depolarizes to a threshold of about −50 millivolts. Once initiated, an action potential propagates non-stop down to the axonal terminal.

of neurotransmitters from the nerve ending into the synaptic cleft, and by the receipt of these messages by the specific receptors on the post-synaptic membrane. Upon receiving the message, the post-synaptic membrane produces a graded resting potential (not action potential)

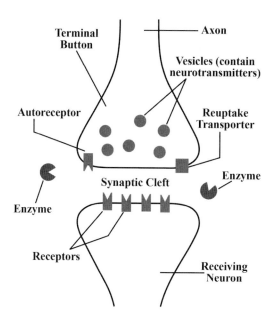

Fig. 7.4. Schematic cross-section of a typical synapse. The upper portion (pre-synaptic component) is the nerve ending; the lower portion (post-synaptic component) is a dendritic spine. Between the two is a narrow space called the synaptic cleft. The flow of information is from the pre-synaptic to the post-synaptic neuron. When a nerve impulse arrives at the axonal terminal, calcium influx triggers the release of neurotransmitters from the vesicles. Four things can happen to the released transmitter. It can (1) hit the post-synaptic receptors for further propagation of message; (2) be degraded by enzymes; (3) be taken up into the pre-synaptic terminal for storage; (4) hit the pre-synaptic auto-receptors for negative feedback. [Courtesy Michael W. King.]

which is one of two types: excitatory post-synaptic potential (*EPSP*) or inhibitory post-synaptic potential (*IPSP*), depending on whether the receptor responds by depolarization (becoming *less* negative inside) or hyperpolarization (becoming *more* negative inside), respectively. When depolarization reaches a threshold it triggers the initiation of an action potential in the axon hillock (origin of the axon),thereby propagating the nerve impulse along the axon toward the terminals. As a rule, membrane depolarization excites a neuron, while hyperpolarization inhibits it.[3]

The simplest neuronal circuit consists of an afferent (sensory) and an efferent (motor) neuron connected by a single synapse, as in a simple reflex arc. Through special receptors in the periphery, a sensory neuron transduces a stimulus (a change in environment) into a nerve impulse and transmits it from the nerve ending toward the soma and then to the central nervous system. Some receptors are sensitive to mechanical forces such as touch, pressure, stretching, and gravity (position sense); others are sensitive to thermal vibration, sound waves, and electromagnetic waves (light); yet others to molecular conformation as in taste and odor. A motor neuron, on the other hand, transmits impulses from the central nervous system to muscles for movement and locomotion, or to glands for hormonal secretion or waste product excretion (such as sweat). A sensory neuron connected directly to a motor neuron forms a monosynaptic arc, as the one involved in the knee jerk reflex. Polysynaptic arcs are formed when one or more neurons (the inter-neurons) interpose between the sensory and motor components. These associative neurons increase in number as the nervous system grows in complexity, so much so that in the human brain most neurons play associative roles. The outcome is an enormously complex communication network.

It should be noted that neurons in the central nervous system do not just serve as a passive conduit for nerve impulses. Many engage in spontaneous background activity in the absence of a stimulus, a phenomenon believed to be related to the state of consciousness, as well as the circadian rhythms of the body.

The message traversing across a synapse is mediated by molecules called neurotransmitters. This chemical relay is much slower compared to the propagation of an action potential. Furthermore, synaptic transmission is not stereotyped, being subject to multiple factors affecting speed and efficiency. The resulting delay across synapses provides endless opportunities for the fine-tuning of inter-neuronal messages.

In a nutshell, synaptic transmission works as follows. When a nerve impulse arrives at the pre-synaptic ending, it triggers an influx of calcium that releases the neurotransmitter from the synaptic vesicle into the synaptic cleft, where the transmitter can act on the post-synaptic receptors (Fig. 7.4). Whether neurotransmission elicits an action potential in the

Table 7.1. Major Neurotransmitters and their Functions

1. Acetylcholine (ACh): Causes skeletal muscle contraction; parasympathetic component of the autonomic nervous system; involved in cognitive function.
2. Norepinephrine (noradrenaline): Sympathetic component of the autonomic nervous system; promotes arousal, vigilance, and aggression; enhances memory.
3. Glutamic acid: Excitatory transmitter; essential for synaptic mechanism of learning and memory consolidation.
4. Gamma amino-butyric acid (GABA): Inhibitory transmitter; counters the excitatory effect of glutamic acid.
5. Dopamine: Controls muscle tone; involved in motivational and reward behavior; over activity leads to hallucination.
6. Serotonin: Responsible for mood stabilization; involved in sleep and feeding behavior.

Note: Norepinephrine, dopamine and serotonin are biogenic amines, since they all contain an amino group. Glutamic acid and GABA are amino acids. Glutamic acid is a natural constituent of protein. For structures of the neurotransmitters, refer to Appendix A.

---

receiving neuron is a matter of mass action, depending on temporal and spatial summation of all the post-synaptic potentials.

There are a handful of major neurotransmitters (see Table 7.1). Acetylcholine causes muscle contraction, and mediates parasympathetic activities in the body and internal organs, while norepinephrine mediates sympathetic activities. (Epinephrine, a variant of norepinephrine, also known as adrenaline, is a stress hormone produced by the adrenal gland when triggered by sympathetic activation.) Inside the brain, both agents play a role in learning and memory, and in maintenance of cognitive function. Glutamic acid and gamma amino-butyric acid (GABA) both function inside the brain, the former as an excitatory and the latter an inhibitory neurotransmitter. Glutamic acid in particular is essential for synaptic learning. Dopamine controls muscle tone by acting on basal ganglia; it also mediates motivation-and-reward behavior. Over-activity of dopamine is correlated with hallucinatory and other schizophrenic symptoms. Serotonin is a mood regulator and is essential for sleep and feeding. The minor neurotransmitters include: aspartic acid, taurine, glycine, histamine, adenosine, nitric oxide, and carbon monoxide, the latter two being gaseous transmitters.

For each transmitter there are several types of receptors with subtle differences in response. Although each neuron produces and releases only one type of transmitter, it can receive multiple incoming transmitters, some with opposing effects. This convergence of influences makes the outcome difficult to predict. Some transmitter receptors are ion channels, allowing the influx of ions such as sodium, calcium, or chloride (ionotropic receptors). Others activate intracellular second messengers (mainly cyclic AMP) through G-protein-coupled receptors (metabotropic receptors).

Neuromodulators are small peptides that do not directly alter post-synaptic electrical potential but indirectly influence the action of transmitters through intracellular second messengers.[4] Many neuromodulators coexist with the major transmitters in the same neuron, and the distinction between the two is not always clear-cut. All in all, there are about thirty neurotransmitters and neuromodulators combined.

Other than neurons, glial cells, which do not conduct action potentials, are present in the nervous system; they comprise 90% of cells in the brain. These are: oligodendrocytes (known as Schwann cells in the peripheral nerves), which, by forming myelin around axons, serve as insulation to speed up nerve conduction; astrocytes, which aid in the metabolism of neurons, in the processing of neurotransmitters, and in the maintenance of the internal environment of the brain by controlling the blood-brain barrier; and microglia, which possess immune and scavenger functions.

7.2 From Nerve Net to Brain: Evolution of the Nervous System

The nervous system increased in complexity as evolution proceeded, from simple nerve nets to ganglia and finally to the brains of higher animals. The following provides a glimpse of this biological procession over hundreds of millions of years.

The simplest form of nervous system appeared in the small fresh water animal called hydra (Fig. 7.5A). Hydra has a tubular body a few

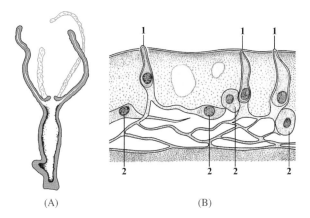

Fig. 7.5. (A) Longitudinal section of a hydra attached to a substrate, with tentacles extending upward surrounding the mouth opening. (B) Fine structure of a hydra showing: (1) sensory cells and (2) epidermal nerve net. [Source: (A) Wikimedia Commons; (B) See Note 32: Hyman LH. (1940); after Groselji. (1909).]

millimeters long with a base attached to any fixed structure in the water. It has a single upward opening, surrounded by tentacles, that serve as both a mouth and an anus. Upon contact with a prey, the tentacles capture it with hook-like structures and bring it into the mouth for digestion. The indigestible residue is subsequently extruded through the same opening. When startled, the hydra retracts into a small ball, a stereotype response irrespective of the direction of the stimulus. Hydra has a nerve net made up of primitive nerve-like cells that conduct a slow action potential (Fig. 7.5B). The cells are interconnected but these connections do not show the full characteristics of a synapse, as there is no one-way transmission, and the impulse appears to diminish in strength along the course. These conducting cells are diffuse and equally distributed so that one part of the nervous system behaves essentially the same as any other part, showing no polarity in message transmission. Communication within the animal body is slow and inefficient. Nonetheless, they do exhibit the phenomenon of sensory adaptation or habituation, the simplest form of learning, in that they cease to capture prey when they are food-sated.

Like the hydra, the free-swimming umbrella-like jellyfish has a radial body plan, but the neurons show discrete synaptic transmission. Instead of a diffusely distributed nerve net, jellyfish neurons tend to group together. Some exhibit endogenous rhythmic activities. Some nerve processes become longer than others and are arranged in parallel bundles for rapid and unidirectional conduction. The nerve nets evolved into nerve rings, which receive messages from well-defined sense organs and discharge into swimming musculature (Fig. 7.6). There are two types of sense organs: one is light sensitive which detects the direction of sunlight, the other responds to gravity with respect to body orientation. Refinement of the neuromuscular system enables a jellyfish to engage in free swimming and procuring food by coordinated movements, consisting of alternate contractions of the circular and longitudinal muscles, expansion of the disc and elongation of the stalk, and swaying movements of the body.

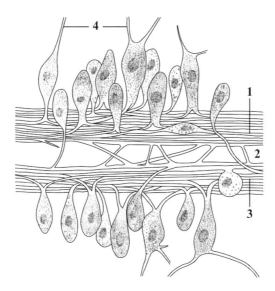

Fig. 7.6. Nerve ring of a jellyfish. (1) Upper nerve; (2) fibers crossing the upper and lower nerves; (3) lower nerve; (4) connecting fibrils to sub-umbrella net. The dome of the jellyfish umbrella is oriented toward the top. [See Note 32: Hyman LH. (1940); after Hyde I. (1902).]

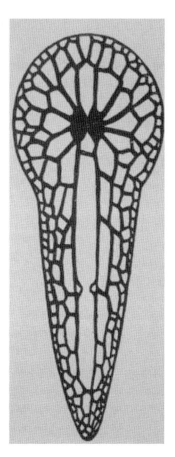

Fig. 7.7. Nervous system of a flatworm, showing a head ganglion (with two lobes) in the front part (top) and multiple smaller ganglia along the body. [See Note 33: Hyman LH.]

Further centralization of the nervous system led to the formation of ganglia, which are aggregates of neurons numbering in hundreds to thousands, as seen in worms (Fig. 7.7). These animals have a bilateral body plan and a longitudinal axis with well-defined front and rear ends. A ganglion is not a simple relay station of sensory and motor neurons. Rather, it serves as an integrative center containing numerous associative neurons. Distributed along the longitudinal axis of the body, ganglia perform semi-autonomous, regional functions, yet are part of a hierarchy, subordinating to a head ganglion at the front end where major sense organs and the oral opening are situated. Thus, ganglionization is accompanied by centralization, encephalization and hierarchization, and results in improved locomotion and feeding activities. In terms of behavior, these animals exhibit not only habituation but also simple conditioning.

As evolution proceeded and the head ganglion increased in size, the brain was formed. In the octopus, the brain has 168 million cells contained in fourteen distinctive lobes. The octopus brain is compartmentalized, allowing a high degree of division of labor. The octopus is unusually clever among invertebrates, possessing a large behavioral repertoire that includes elaborate posturing, sexual display and copulation; they have good visual acuity and are capable of maze learning, home construction, and territorial defense.

Insects are the other invertebrates that have a well-formed nervous system that includes a primitive brain. They have good visual, olfactory (smell) and tactile (touch) senses. Their compound eyes are good at detecting flickers of light. Their sense of smell matches the best seen in vertebrates. Their sense of touch makes possible a meticulous exploration of the environment. A good position sense enables them to orient their body. These, combined with their articulated appendages with reciprocal innervations, endow them with agility of movement and rapidity of response, greatly enhancing the chance of survival in a dangerous environment.

7.3 Nervous System of the Vertebrates

With the advent of vertebrates in the evolutionary scene, one sees enormous changes in the nervous system, most of which occurred in the brain. There was over a thousand-fold explosion in brain size, from a centimeter in fish to 15 cm in man, and from a few grams to one and a half kilograms. Commensurate with this is the great increase in complexity of the neuronal network. Some features of the vertebrate brain and how they evolved are shown in Figs. 7.8 and 7.9.

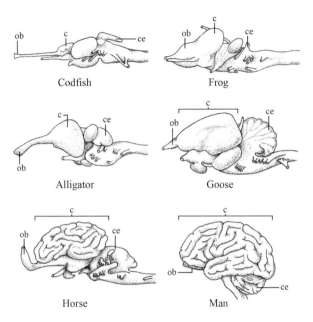

Fig. 7.8. Vertebrate brains in evolution: (c) cerebrum (brain proper); (ce) cerebellum (for balance); (ob) olfactory bulb (for smell). Drawings not to scale. Note the relative expansion of the cerebrum and the shrinking of olfactory bulb in the human brain. [See Note 34: Romer, Parsons; permission Cengage Learning.]

The brain of a vertebrate is encased in a protective bony cavity. This provides the brain, basically a soft, gelatinous mass of matter vulnerable to physical damage, a chance to develop delicate structures necessary for higher mental functions. But the skull also sets an upper limit for the size of the brain to evolve. The maximum skull size, on the other hand, is constrained by the pelvic opening of the mother through which a newborn must be delivered. The brain may appear left-right symmetrical, but in reality some division of labor exists. For example, most people have their language center located on the left side.

The spinal cord is the downward extension of the brain that goes through the entire length of the body. It comprises two major systems, the ascending tracts which convey sensory information from the body to the brain, and the descending tracts made up of motor neurons that

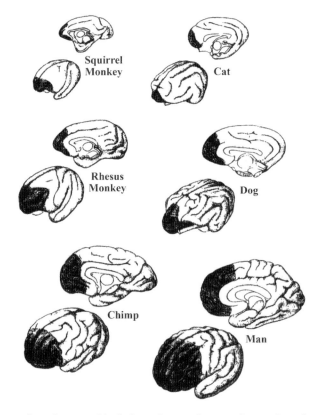

Fig. 7.9. Prefrontal cortex (shaded area) in evolution. The prefrontal cortex is the part of the frontal lobe minus the motor and speech areas, and is known for performing executive functions. Orientation of the brains is toward the left. Drawings not to scale. [See Note 35; permission Oxford Univ. Press.]

execute movements on command from the brain. The spinal cord is also encased in a bony structure called the vertebral column. It is important to remember that most of the fibers in the cord cross over the midline on their way to and from the brain, so that sensory fibers from the right side of the body end up on the left side of the brain; motor fibers, likewise, activate muscles on the opposite side.

For convenience of study, the vertebrate brain is divided into three sections: (1) the hindbrain, (2) the midbrain, and (3) the forebrain (Fig. 7.10). The hindbrain comprises the medulla, pons and cerebellum,

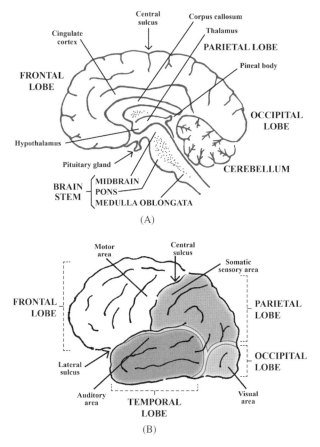

Fig. 7.10. Human brain in medial (A) and lateral (B) views. "A" is the right cerebral hemisphere viewed from the midline of the brain; "B" is the left hemisphere viewed from outside of the brain. The orientation in both cases is such that the head (front end) is pointing to the left. The frontal lobe occupies about 30% of the cerebrum. In "A" the cerebrum, brain stem, and cerebellum are shown; in "B" only the cerebrum is shown. The major parts of the cerebrum are: frontal lobe, parietal lobe, occipital lobe, and temporal lobe.

the latter being attached to the pons. The medulla regulates all the vital functions of an animal, such as heartbeat, blood pressure, and respiration. The cerebellum is mainly for balance and coordination. The midbrain is a transition area between the hindbrain and the forebrain. The medulla, pons, and midbrain are collectively referred to as the "brain stem." An animal with the forebrain separated from the brain stem loses consciousness but can survive for some time.

Structures anterior to the midbrain are collectively called the forebrain, commonly known as the *cerebrum* or the "brain" proper in higher animals. It is made up of a mantle of gray matter (cerebral cortex) and a deeper part of the brain called white matter. The gray matter comprises neuronal cell bodies and their unmyelinated processes, whereas the white matter consists mainly of nerve tracts of myelinated axons. Embedded in the white matter are pockets of gray matter (called "nuclei"), making up of neuronal cell bodies, that include such structures as the amygdala, basal ganglia, and thalamus. The cerebral cortex is a latecomer in evolution and correlates with the appearance of intelligence. It is architectonically arranged in layers. In lower vertebrates such as fish, the cortex has only three layers of neurons, and is referred to as the allocortex or archicortex, whereas in higher animals including man it has six layers and is called the neocortex.[5] Starting from the amphibians, an olfactory cortex for the sense of smell came into being, which became relatively insignificant in primates as the sense of smell lost its importance. During evolution, a rudimentary neocortex first appeared in reptiles and became increasingly important thereafter. In the human brain the neocortex occupies over 90% of the cerebral cortex. It grows so much in bulk that it folds inward upon itself (to increase the surface area), forming gyri (ridges) and sulci (furrows), giving the brain surface a peculiar wrinkled and bumpy look. The human cerebral cortex is topographically differentiated, with areas dedicated to such functions as vision, hearing, touch, movement, and a large area for associative and integrative purposes. The prefrontal area (the region in front of the motor and pre-motor areas) is particularly important in humans, as it is believed to be the seat of rational thinking and executive function, the derangement of which has been implicated in schizophrenia.

Attention should also be paid to the area called the *limbic system*, best seen from the medial side of the brain, consisting of the older cortex and its adjacent structures surrounding the brain stem (see Fig. 7.11).

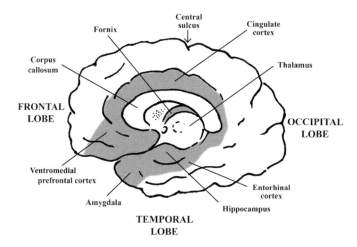

Fig. 7.11. The limbic system (shaded areas) of the human brain (medial view of the right hemisphere with the head pointing left). The following limbic structures form a ring around the brain stem: ventromedial prefrontal cortex, cingulate cortex, entorhinal cortex, hippocampus, and amygdala (deep in the brain). Other structures are labeled for landmark identification.

It is believed to be important for primitive drives and motivation, and for this reason has been referred to as the "visceral brain."

The autonomic nervous system, which consists of a chain of ganglia (carrying messages originating in the hypothalamus) alongside the spinal cord, is responsible for energy consumption during emergency (sympathetic) and for energy conservation at rest (parasympathetic). The autonomic nervous system will be elaborated in *Chapter 9* in connection with emotion.

7.4 Synaptic Plasticity: The Key to Learning

Synaptic-based neuronal circuitry is the foundation of the Hebbian theory of learning. Hebb, in the 1940s, suggested that repeated firing of one neuron increases the synaptic efficacy with the next.[6] Perhaps the strongest evidence in support of synaptic learning is the phenomenon of long-term potentiation, a topic to be taken up in *Chapter 10: Self and Memory*.

Although most adult neurons do not divide, the synaptic connections are constantly in a state of flux. High-resolution microscopy in the living, adult mouse cerebral cortex reveals the dendritic spines in constant morphologic changes and movements, observable in a span of less than ten minutes.[7]

Plasticity of the developing brain is evident in its remarkable ability to recover from injury. Newborn rats with almost half a cerebral hemisphere removed may grow up with no apparent motor deficit. Humans born with extensive injury to the language centers in the left hemisphere grow up with little speech impairment (confirmed by functional MRI). In these instances, the opposite hemisphere takes over the lost function as a result of circuit rewiring. However, the regenerative capacity decreases rapidly after a critical period. For instance, human infants with congenital cataract, if uncorrected within the first three years, remain blind even after surgery in later life.

Molecular recognition in synaptic connection is most dramatically demonstrated in lower vertebrates. Roger Sperry, in the 1940s, showed that when the optic nerve of a frog is cut and allowed to regenerate, the individual axons at the cut end from the eye seek out their respective partners in the midbrain in a topographically precise manner. The matching of the two parts by chemical recognition is known as chemo-affinity.[8]

In recent years, molecular markers on neuronal surface carrying repulsive or adhesive messages have been confirmed. In the fruit fly, a gene called DSCAM encodes a large family of axon guidance receptors. Alternative splicing of the gene gives rise to 38 thousand different cell surface recognition proteins.[9] Other DSCAM proteins are found in mouse,[10] chick retina,[11] and the invertebrate sea slug.[12]

7.5 Parallel Evolution of Brain and Behavior

Figure 7.12 shows changes in adaptive behavior with respect to phylogeny. All behaviors are outcomes of interplay between internal (genetic and epigenetic programs) and external factors (environment) shaped by natural selection. When the "brain" is nothing but a head ganglion at the

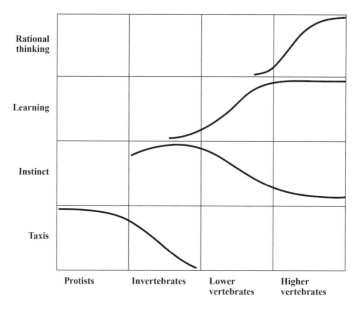

Fig. 7.12. Diagram showing evolution of behavior. The mode of behavior goes from the simplest to the most complex from bottom to top. Time course of evolution goes from left to right. Values are not strictly quantitative. Note that instincts, defined as genetically determined behaviors, never disappear from higher vertebrates, including humans.

front end of a flatworm, its function is no more than a hub for sensory and motor activities, not far from a reflex center. As the head ganglion expands in size, as in insects, instinctive motivational behaviors become evident. Some sort of primitive "urge" seems to be in place when a fly is attracted to a sugar solution.[13] Lower vertebrates express strong emotions in feeding and mating. Mammals develop a much more elaborate brain structure, in particular the neocortex, enabling them not only to better coordinate their behavior but also to store information for associative and retrieval purposes. The development of the neocortex culminates in the appearance of the prefrontal lobe, which in humans occupies one third of the entire cerebral cortex. The advent of the prefrontal cortex enhances rational thinking, enabling an animal to "foresee" the consequences of alternative behaviors in order to take proper action. Rational thinking also helps to put a brake on the urge from the primitive emotional centers. It is the balance of reasoning and emotion that leads to a normal life. Basic instincts, however, never totally disappear, remaining an undercurrent even in human behavior. Thus, we see the impulsive craving for food of an infant replaced by the well-controlled table manners of an adult, both driven by the need to feed. Nevertheless, under stress or in brain damage, the primitive impulse may resurface and adult behavior can revert to that of a baby.

7.6 Examples of Behavior at Different Levels of Neural Complexity

Following are some examples of behavior in phylogenetic order:

7.6.1 *Light-avoidance in planarians*

The flatworm planarian displays a light–avoidance behavior called negative phototaxis. The central nervous system consists of a bi-lobed primitive brain (the head ganglion) and two ventral nerve cords (Fig. 7.7). A pair of light-sensitive organs at the front end are connected to the brain. Planarians normally exhibit light avoidance behavior by sending signal from the eyes to the brain. After decapitation, the brain and "eyes" regenerate and phototaxis is resumed.[14]

7.6.2 *Switching light preference in fruit flies*

The fruit fly *Drosophila* prefers darkness in the early larval stage but prefers light in adulthood. The switch from photophobic to photophilic behavior is controlled by two pairs of neurons downstream from the photoreceptors. The change enables the adjustment of the animal's response strategy to environmental stimuli according to biological needs.[15]

7.6.3 *Instinctive behavior in insects and higher animals*

Social insects are equipped with a rich repertoire of unlearned behaviors, especially with respect to intra-species communication. Honeybees use a unique "dance" language to convey information about the direction and distance of the source of nectar. Ants are able to communicate with pheromones, chemicals they excrete that provoke a stereotyped response from other ants. Insects also use a combination of inborn ability and learning. Thus, honeybees employ the sun and polarized light in their navigation, while using their internal clock to estimate the change in sun angle. Ants have been shown to learn to find food by going through a maze, after repeated trial and error.

Shore-dwelling sand fleas have a propensity to jump to the sea along the coastline. When carried inland, those living on the west cost of Italy will jump to the west, in the direction of the ocean. Interestingly enough, when transported to the east coast, the same animals still jump to the west, in this instance away from the ocean, jeopardizing their survival.[16]

One of the most intriguing phenomena in nature has been the migration of North American monarch butterflies. Each year these insects fly 4,000 kilometers from Canada to winter in Mexico, where they mate, and return when the weather gets warm. Not only are they able to land in the same spots year after year, the time needed for a round-trip journey exceeds the lifetime of a butterfly. It takes one generation to fly south and three to fly north. How the geographic information is encoded and transmitted over multi-generations remains a mystery. Most likely the process involves epigenetics as well as genetics, along with environmental cues such as the Earth's magnetic field and the orientation of sunlight.[17]

During mating time each female squid lays thousands of egg-capsules. Upon contact with these egg masses, a male squid's behavior changes abruptly, becoming extremely aggressive and ready to fight with other males to pair with the fertile female. Thus, the most dominant male gains the greatest number of copulations, and females benefit by obtaining sperm from the most vigorous competitor. The active factor in the egg capsules is a protein contact-pheromone termed Loligo β-MSP. The synthetic recombinant protein alone (in the absence of the eggs or the females) is capable of inducing similar male behavior.[18]

Loggerhead turtle hatchlings use Earth's magnetism for global positioning to navigate long-distance to the favored part of the Atlantic Ocean. They appear to be able to locate the correct destination with both the longitudinal and latitudinal coordinates of the Earth's magnetic map.[19]

Young ducks and geese have an instinctive fear of their natural predators. Figure 7.13 shows a model that has the shape of a flying bird. When the model sails to the right, simulating a predator, the young birds go into panic and try to escape; but when it sails to the left, simulating a goose, they do not respond. No learning is involved.

7.6.4 *Imprinting in young animals*

Imprinting is a primitive type of learning that borders on instincts. It is acquired at a very young age, within a critical period, whose effect may last a lifetime. One well-known example is the "following behavior" of goslings reported by Konrad Lorenz. Goslings are imprinted to follow the first moving object they see after hatching, whatever it is, even preferring it to their real mother. Another example is the upstream homing

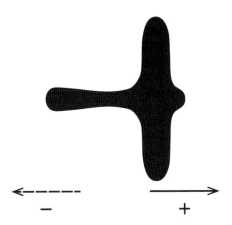

Fig. 7.13. Instinctive response of young birds to flying objects. Ducklings or goslings respond with fear when the object (or shadow) moves to the right, simulating a hawk, but not when it moves to the left, simulating a goose. No experience is necessary. [See Note 36; courtesy Oxford Univ. Press.]

of salmon. Salmon spawn in rivers high up in the mountains and subsequently migrate to the sea. When mature and ready to mate, they swim upstream to their birthplace. It was demonstrated that the cue for homing is the odor of the water of their birthplace imprinted on the young fish.

7.6.5 *Trial and error versus reasoning*

Higher vertebrates are capable of learning from experience. Figure 7.14 shows a raccoon trying to reach a food by random trial. After several unsuccessful attempts, it learns from experience that it must first move away from the food source in order to get it. Figure 7.15 shows a chimpanzee solving the problem of getting a banana that is hung from the ceiling without resorting to trial and error; it reasons that the banana can be reached by stacking boxes on top of one another.

It is worth mentioning that birds of the corvid family, which includes ravens, crows, jays, magpies, and nutcrackers, have high intelligence and good reasoning power. They can fashion tools or carry out

Fig. 7.14. A raccoon that is unable to reach a bait, not knowing that it must first move away from the food and turn around the hitch to get it. The attempt is successful after trial and error. [See Note 37; courtesy Dover Publications.]

Fig. 7.15. A chimpanzee getting banana by stacking boxes. It does this by reasoning and succeeds on the first try. [See Note 37; courtesy Dover Publications.]

elaborate procedures to retrieve food. Food-storing birds can remember the numerous sites where they hide their food. What is more interesting is that these birds steal food from one another. For example, if bird A watches bird B hiding its food, bird A will steal it while B is not around. However, if bird B notices that it has been watched by bird A when hiding the food, B will guard against the encroachment of A, or move the food to a different place before A has a chance to steal it. Meanwhile, bird B will ignore the approach of another bird (C), which did not know the whereabouts of the food. The ability to recognize the mind of other individuals and to anticipate their future actions puts the intelligence of these birds on par with the great apes, despite the birds' small brains.[20]

7.6.6 Language function

Language is the pinnacle of achievement of intelligent animals. Although other animals are capable of rudimentary language (dogs can respond to a large vocabulary of single words, and chimpanzees can form simple sentences), only humans can formulate complex statements out of conceptualization. Humans have the unique ability to manipulate and organize symbols in a recursive and nested manner — one idea sits in another, which in turn sits in yet another, *ad infinitum* (recursively enumerable).

7.7 Basic Features of a Well-developed Brain

A well-developed brain has the following characteristics: (1) It has a *hierarchical* organization, wherein members of one rank are nested in, and subordinate to, a member of the next higher rank. (2) It has a *modular* structure, in that each component can serve as a member of several functional units. (3) Neuronal circuits, formed by synaptic connections, are malleable and self-adjusting. Multiple circuits are connected into networks according to the principles of hierarchy and modularity.[21] Figure 7.16 shows the complexity of neuronal circuits and networks. (4) Neuronal circuitry appears to show redundancy in function. In other words, some neurons are put on reserve, ready to be called into action in times of need. For instance, any loss of neurons has to exceed a certain threshold for a deficit to be detectable. In Parkinson's disease, symptoms are not apparent until 80-90% of neurons of the *substantial nigra* in the midbrain are gone. (5) The brain is capable of *information processing*. It follows a certain algorithm and is capable of parallel processing and reentrant dynamics. Except for speed, the brain is far more superior to any computer known to man, being capable of self-programming, self-adjustment, and self-correction. Despite its hierarchical organization, there is no evidence that a single neuron serves as a supreme ruler of the entire brain. For this reason, it has been compared to an orchestra without a conductor.[22] (6) Most importantly, unlike most electronic

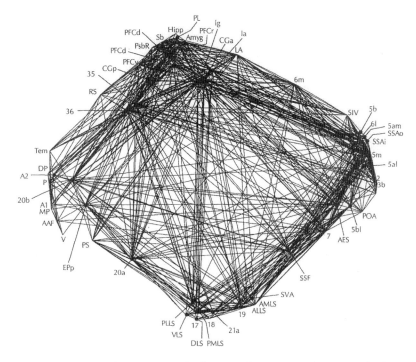

Fig. 7.16. Diagram showing network of neuronal circuitry in the cat cerebral cortex. The picture represents 64 areas with 1,134 connections between them. The map depicts only their connections but not their anatomical locations. The names of different areas are presented in abbreviations but they are irrelevant to the current discussion. [See Note 38; permission Cell Press.]

computers, the brain does not run in a strictly logical manner; it is tainted by emotions and feelings made possible only by the complex cellular and chemical nature of the nervous system. This deviation from perfect logicality with an occasional outbreak of capriciousness is perhaps what makes a human being human.

7.8 Hierarchical Organization Leads to Stepwise Abstraction

Hierarchical organization of neurons in the brain permits increasing degrees of abstraction in cognitive function, whereby particular, detailed inputs from the outside world are filtered to extract the salient and generalized features — from concrete sensation to structured perception to conceptual idealization. Take an example from vision. The first step is the sensory data of brightness, colors, shapes, texture, etc., arriving in the occipital primary visual cortex. The second step takes place in the inferior temporal cortex where individual views of a face from different angles are integrated to represent the face of a person. Finally, the general concept of a given person (say, Jennifer Aniston, a movie star) is encoded in the entorhinal cortex and hippocampus of the medial temporal lobe, where a special group of neurons called *concept cells* are activated. Concept cells form an invariant response to multimodal sensory input, attending only to the relevance (meaning) of the stimulus while leaving out the particulars. The same "Jennifer Aniston neuron" responds to different pictures of her in multiple poses, and to her name in writing and in spoken languages as well. The neural mechanism of this stepwise abstraction has been demonstrated by single cell recording in live persons. It is believed that concept cells do not act in isolation. Rather, each concept is assembled in a network that overlaps with networks of other concepts, building different degrees of association, a process important for episodic memory.[23] We can speculate that neuronal assemblies representing increasing degrees of abstraction — covering such items as gender (female), professional group (actress), ethnicity (Caucasian), and even humanity — could be present. The hierarchical nature of neuronal networks makes possible the stepwise abstraction of information.

7.9 Singularity of *Self* and Oneness of Psychic Experience

When I wake up in the morning, even before opening my eyes, I feel like I am one person. This feeling of *self* continues throughout the day — and my whole life. Even when I am dreaming, I never experience multiple *selves*. In fact, *self* is always one, never two thirds, or one and a half. When Penfield stimulated the temporal cortex of his epileptic patients to create an aura, the patients always reported one experience at a time.[24] In fact, the singularity of *self* is the most exquisite product of our brain. The nervous system is so intricately coordinated and integrated that the whole functions as a harmonious "one," or gives the subjective feeling of one. The singularity of mind is taken for granted, and from there comes the religious tenet of "one body, one soul."

To many people's surprise, this elegance of oneness is true only for healthy brains, but not for some abnormal brains. To explain this, let me start from the "bilaterality" of our body scheme, an inheritance from our distant ancestors in the Cambrian period, about 500 million years ago, when we were in the ranks of the trilobites. Like most animals, our body has a left-right symmetry, which is roughly but not exactly true (the limbs and kidneys are represented equally on both sides, but the heart is single and toward the left). The brain is also composed of two parts (the left and the right hemispheres), similar in appearance but unequal in fine-tuning. The left cerebral hemisphere has a propensity for language, reasoning, calculation, abstract thinking, whereas the right is good for space orientation, emotion and intuition. For some unclear reasons, extensive damage to the right cerebral hemisphere produces a left-sided neglect, but a left hemispheric injury does not lead to a reciprocal deficit (Fig. 7.17).[25]

The most dramatic fragmentation of *self* happens when the two cerebral hemispheres are disconnected surgically. This procedure, by which the corpus callosum (a massive fiber tract connecting the two hemispheres) is cut as a treatment for intractable epilepsy, produces a person who is normal in terms of gross sensations and movements, but lacks communication between the two hemispheres, so that knowledge acquired by one is no longer shared with the other. In the 1960s, Roger Sperry and his colleagues conducted detailed psycho-behavioral analysis of these "split-brain" patients and found that, in certain instances, they act like a person having two minds, sometimes even conflicting minds. For example, since language function is performed only in the left hemisphere, the patient can verbally report things that the left hemisphere

Fig. 7.17. Demonstration of left-sided neglect in a patient with right hemisphere damage. The patient was asked to copy a group of pictures (on the left). Note that the patient's copies (on the right) lack details of the left half for each object. This phenomenon is unrelated to visual loss.

knows about. Conversely, when the stimulus is presented in such a way that the information enters only the right hemisphere, the patient cannot talk about it, but he can command the left hand to respond in a non-verbal manner, such as reaching and touching the object.[26] The fact that a split-brain person has a "divided mind" strongly supports the contention that mind is grounded in matter — the brain — and dispels the possibility that the mind is capable of independent existence. [27]

7.10 The Brain and the Making of the Human *Self*

As I emphasized earlier, *self* is a *system* and not a molecule, a group of cells, or an organ — not even the brain. Nonetheless, a well-developed nervous system is necessary for the full expression of *self*. It is through the mechanism of the brain that *self* assumes the capacity of consciousness, first of the world, and finally of itself.

Sigmund Freud divided human *self* into *id, ego and superego*, corresponding, respectively, to animal impulse, the day-in-and-day-out *self*, and the moral and idealized *self*.[28] Kandel suggested, and I agree, that the three levels of *self* can be roughly correlated with the phylogenetic stratifications of the brain: *id* is the function of the lowermost part — the brain stem and the limbic system; above this are the cerebral hemispheres which give rise to *ego*; whereas *superego* is the function of the upper- and outer-most part of the hemisphere — a thin layer of gray matter called cerebral cortex, especially the prefrontal cortex.[29] The prefrontal cortex occupies the highest level in the hierarchy of the nervous system and its function is what most distinguishes humans from other animals. It is involved in planning, foresight, self-control, self-restraint, and deferred gratification. Figure 7.18 is a diagram of this scheme.

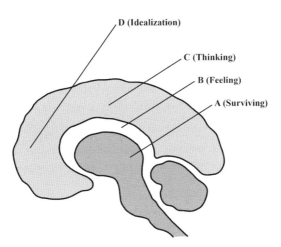

Fig. 7.18. Evolutionary stratification of the brain in the building of the human *self*. The three layers conceived by Paul MacLean in the 1960s are, approximately: (A) the archipallium for survival; (B) the paleopallium for feeling and emotion; (C) the neopallium for rational thinking. When translated into psychoanalytic terms, *id* corresponds to (A) and (B); *ego* corresponds to (C), while *superego* resides in a special part of the cerebral cortex equivalent to the prefrontal cortex (D). [See Note 39.]

In the next four chapters, the important mental functions will be examined in detail with reference to the various aspects of the human *self*. Emphasis will be placed on consciousness, emotion, memory, and free will. Their survival values will be discussed.

7.11 The Brain and the Outside World

The brain cannot perceive itself; it can only perceive something else — a body part, another person, or the environment. An analogy is with a *mirror*. A mirror can never reflect itself; it can only reflect other things. Even if we put a second mirror in front of the first, what the second mirror reflects is what the first reflects (plus, of course, the edges of the first mirror). A mirror alone has nothing to be reflected. Likewise, whatever the brain does has to be referred to something outside of it. When a person thinks, he has to think of something, however "abstract" the subject can be. No one can think of a vacuum or a void. Even when you think of the word "vacuum," certain mental pictures come to mind — a page in a physics textbook with that heading, a sealed cylinder in a laboratory attached to a pump, or a star-studded outer space. How about mathematical thinking? Sure enough, we have mental pictures of geometric forms (a triangle, a cylinder, a parabolic curve), or digits on our mental blackboard. When we discuss politics, we envision voting, inaugurations, bomb explosions, assassinations, or lengthy speeches of demagogues.

On the other hand, the brain also acts like a *projector*. It projects the contents of its activity (intentions, actions, feelings) outward. An emotionally aroused person feels tension in the muscles, palpitation in the heart, and contraction in the stomach, but nothing in the brain. When suffering from a headache, the pain is in the scalp, never in the neurons. Likewise, stimulating the motor cortex leads to limb movement, not brain movement. In the condition called thalamic syndrome, lesion in the thalamus causes intolerable pain in the extremities, but not in the thalamus. In extreme cases, the brain can even project sensations to non-existing body parts, as in "phantom limb" syndrome, frequently seen in patients with an amputated limb. It was reported that, in an experiment in which the brain structure amygdala was electrically stimulated, a sensation of nausea and belching was felt in the stomach, although no real gastric movement could be detected.[30]

7.12 Limitations Imposed by the Physical Brain

It is doubtful that the human brain will evolve into an ever more powerful thinking machine. Conventional wisdom says that the pelvic opening of the mother imposes a limit to the size of the baby's head. But even if this difficulty is overcome (say, by routine caesarian section or by enlarging the cavity), basic physical laws provide an insurmountable constraint to the structure of the neurons. The following are points to consider: (1) Miniaturization of neurons can theoretically pack more cells per unit volume. But, thermal vibration renders small-diameter axons inherently unreliable for ion movement across membrane (the way action potentials are made), reducing reliability of nerve signals. (2) Enlarging axonal diameter can speed up nerve conduction, but the expenditure of energy outpaces the gain in speed, rendering it cost-ineffective. (Current mammalian neurons have already gained speed by wrapping the axons with myelin, an effective insulator.) (3) Enlarging the brain size can increase neuron number, but this requires longer axons for longer-range transmission of message, resulting in sluggish response to the environment, as seen in elephants. Thus, improving the *neuron-based* intelligence hits a roadblock. Fortunately, only humans, of all animals, break through the barrier by inventing extracorporeal information-processing mechanisms, such as language, writing, books, libraries, and, best of all, computers.[31]

Notes and References

1. Kety SS. (1960) A biologist examines the mind and behavior. *Science* **132:** 1861–1870.
2. The speed of a good metallic conductor approaches that of light (3×10^8 meters per second).

3. A post-synaptic potential may or may not lead to an action potential, but an action potential, once formed, will self-propagate to the terminal of the axon.
4. The common neuromodulators are somatostatin, cholecystokinin (CCK), oxytocin, vasopressin, vasoactive intestinal peptide (VIP), Substance P, endorphin, neuropeptide Y (NPY), neurotensin, bradykinin, angiotensin, and thyrotropin-releasing hormone (TRH). The non-peptide neuromodulators include the endocannabinoids (lipids) and hydrogen sulfide (gas).
5. One exception is the hippocampus, which retains the 3-layer structure of a primitive cortex. This suggests that the hippocampus is evolutionarily old.
6. Hebb HB. (1949) *The Organization of Behavior.* John Wiley.
7. Berning S, Willig KI, Steffens H, *et al*. (2012) Nanoscopy in a living mouse brain. *Science* **335:** 551.
8. Sperry RW. (1963) Chemoaffinity in the orderly growth of nerve fiber patterns and connections. *Proc Natl Acad Sci USA* **50:** 703–710; Meyer RL. (1998) Roger Sperry and his chemoaffinity hypothesis. *Neuropsychologia* **36:** 957–980.
9. Hattori D, Millard SS, Wojtowicz WM, Zipursky SL. (2008) Dscam-mediated cell recognition regulates neural circuit formation. *Ann Rev Cell Dev Biol* **24:** 597–620.
10. Fuerst PG, Bruce F, Tian M, *et al*. (2009) DSCAM and DSCAML1 function in self-avoidance in multiple cell types in the developing mouse retina. *Neuron* **64:** 484–497.
11. Yamagata M, Sanes JR. (2008) Dscam and Sidekick proteins direct lamina-specific synaptic connections in vertebrate retina. *Nature* **451:** 465–469.
12. Li HL, Huang BS, Vishwasrao H, *et al*. (2009) Dscam mediates remodeling of glutamate receptors in Aplysia during *de novo* and learning-related synapse formation. *Neuron* **61:** 527–540.
13. Caution should be used when denoting a behavior as instinct. There is a gray area between the learned and the inborn. What appears as instinct may have a learned component upon scrutiny, such as *imprinting* in young animals, which usually occurs at a critical period.
14. Inoue T, Kumamoto H, Okamoto K, *et al*. (2004) Morphological and functional recovery of the planarian photosensing system during head regeneration. *Zoolog Sci* **21:** 275–83.
15. Gong Z, Liu J, Guo C, *et al*. (2010) Two pairs of neurons in the central brain control Drosophila innate light preference. *Science* **330:** 499–502.

16. Dethier VG, Stellar E. (1070) *Animal Behavior.* 3rd ed. Prentice-Hall, New Jersey.
17. Guerra PA, Gegear RJ, Reppert SM. (2014) A magnetic compass aids monarch butterfly migration. *Nature Communications* doi:10.1038/ncomms5164.
18. Cummins SF, Boal JG, Buresch KC, et al. (2011) Extreme aggression in male squid induced by b-MSP-like pheromone. *Current Biol* **21:** 322–327.
19. Putman NF, Endres CS, Lohmann CMF, Lohmann KJ. (2011) Longitudinal perception and biocoordinate magnetic maps in sea turtles. *Current Biol* **21:** 463–466.
20. Heinrich B, Bugnyar T. (April 2007) Just how smart are ravens? *Scientific Am* **296:** 64–71; Raby CR, Alexis DM, Dickinson A, Clayton NS. (2007) Planning for the future by Western Scrub-Jays. *Nature* **445:** 919–921.
21. The importance of circuitry has been borne out by experimental data obtained from the living brain. In a study using fruit fly larvae involving over a thousand identified neurons, it was found that multiple behaviors can be associated with activation of the same neurons, and multiple neurons can trigger the same behavior, demonstrating that pattern and circuitry rather than individual neurons are better correlates of behavior. See: O'Leary T, Marder E. (2014) Mapping neural activation onto behavior in an entire animal. *Science* **344:** 372–373; Vogelstein JT, Park Y, Ohyama T, et al. (2014) Discovery of brainwide neural-behavioral maps via multiscale unsupervised structure learning. *Science* **344:** 386–392.
22. Edelman GM, Tononi G. (2000) *A Universe of Consciousness.* Basic Books, New York.
23. Quian Quiroga R. (2012) Concept cells: The building blocks of declarative memory functions. *Nature Rev Neurosci* **13:** 587–597.
24. Penfield W, Jasper H. (1954) *Epilepsy and the Functional Anatomy of the Human Brain.* Little, Brown, Boston.
25. Nerve fibers from one side of the body (and face) cross to the opposite side before ending in the cerebral cortex. The left-sided neglect includes the loss of attention to the left half of the patient's own body and the left half of his/her environment; this symptom is not explainable by the loss of sensory input.
26. Gazzaniga MS. (1972) One brain — two minds. *Am Scientist* **60:** 311–317.

27. A partial split-brain syndrome occurs naturally as a consequence of a certain type of stroke, in which a lesion is present in the left occipital visual cortex along with the adjacent posterior end of the corpus callosum. In this manner the reading center (present only in the left hemisphere) cannot receive written image from the visual cortex of either hemisphere, but the patient is able to put out written words from the reading center to the motor system of the brain, resulting in the strange condition called "alexia (inability to read) without agraphia (inability to write)." A concrete example is when a person can write but cannot read his own writing a few minutes later. The alexia, however, can be overcome by tracing the letters with a finger, in this instance using the motor system and its sensory feedback as an incoming channel for the words. See: Cuomo J, Flaster M, Biller J. (2014) Right Brain: A reading specialist with alexia without agraphia: Teacher interrupted. *Neurology* **82 (1)**: e5–7. DOI: 10.1212/01.wnl.0000438218.39061.93.
28. Freud S. (1923) *The Ego and the ID*. W. W. Norton.
29. Kandel ER. (2012) *The Age of Insight*. Random House, New York.
30. Van Buren JM. (1963) The abdominal aura: A study of abdominal sensation occurring in epilepsy and produced by depth stimulation. *Electroencephalogr Clin Neurophysiol* **15:** 1–19.
31. Fox D. (July 2011) The limits of intelligence. *Scientific Am* **305**: 36–43.
32. Hyman LH. (1940) *The Invertebrates*. vol. 1, McGraw-Hill, New York.
33. Hyman LH. (1951) *The Invertebrates* vol. 2, McGraw-Hill, New York.
34. Romer AS, Parsons TS. (1986) *The Vertebrate Body*. 6th ed. Saunders, Philadelphia; Brooks/Cole; Cengage Learning, Inc.
35. Fuster JM. (2003) *Cortex and Mind*. Oxford Univ. Press, Oxford.
36. Tinbergen N. (1989) *The Study of Instinct*. Oxford Univ. Press, Oxford.
37. Maier NRF, Schneirla TC. (1964) *Principles of Animal Psychology*. Dover, New York.
38. Scammell JW, Young MP. (1993) The connectional organization of neural systems in the cat cerebral cortex. *Current Biol* **3**: 191–200.
39. MacLean PD. (1973) A triune concept of the brain and behavior. In: Boag TJ, Campbell D. eds. *The Hincks Memorial Lectures*. Univ. of Toronto Press, pp. 6–66; MacLean PD. (1990) *The Triune Brain in Evolution: Role in Paleocerebral Functions*. Plenum Press, New York.

Chapter 8: *Self* and Conscious Experience

Consciousness is your momentary present. As soon as you pass it, it piles up in the rearview mirror as memory.

Overview: *Consciousness is the baseline from which all other mental functions arise. It starts as the awareness of the environment; from there it develops into the awareness of one's own self, separating the "I" from the rest of the world. The highest function of the mind is to reflect on one's own mind, or to be conscious of one's own consciousness.*

Various states of consciousness — alertness, drowsiness, sleepiness, dreaming, and coma — are correlated with electrical activity recorded from the brain. Inside the brain, two systems are involved in consciousness: (1) a discrete system responsible for the perception of specific stimuli, such as seeing a red flower or hearing the beating of a drum; (2) a diffuse system that raises general alertness without referring to any particular object.

8.1 Consciousness Polarizes *Self* and Non-*self*

Consciousness is the ground substance of all mental activities. The emergence of consciousness marks the transition of *self* from the realm of physics and chemistry to the domain of society and humanity. In the cognitive sense, consciousness leads to the separation of *self* from the rest of the world and eventually to the reflection of one's own *self* — the reflective self.

The changing perspectives of *self* during development are portrayed in Table 8.1. When a child is born, there is no distinction

■ 163

Table 8.1. Separation of *Self* from the Outside World

Stage	Age	*Self*	World	Examples
0	Newborn	Nebulous chaos		
1	Infant (3–12 mon)	Perceives physical *self* (interoception)	Perceives physical world (exteroception)	World as object; other people as object
2	Toddler (2–3 yr.)	Reflects on physical *self*	Projects *self* outward	*Self* as object; self-recognition in mirror; self-recognition in picture
3	Young child (4–6 yr.)	Reflects on mental *self* (mind)	Recognizes other people's minds as well as own mind	Passes "theory-of-mind" test

Stage 1: First order perception.

Stage 2: Second order perception; first order reflection.

Stage 3: Third order perception; second order reflection; reflection on reflection; introspection and recursive thinking on *self*.

between *self* and non-*self*. He or she is confronted with a chaotic mix of sense data.[1] The jumble of sensations leads to perception once they are filtered, selected, and categorized. In this manner the first reality of the outside world comes as the taste of the milk, the texture and warmth of the mother's breast, and the tender cooing sound from the parents. These perceptions are called *exteroception*, as they come through the external sense organs — eyes, ears, nose, mouth, and skin.[2] At the same time, a different set of perceptions, called *interoception*, arrives through private sensation of one's own body including the internal organs. The distinction between intero- and extero-ception in Stage 1 leads to the separation of *self* from non-*self*, making possible the exploitation of the latter for the benefit of the former. This is seen when an infant strongly sucks the mother's nipple to satisfy his hungry stomach. When a one-year-old sticks everything into his mouth, he is evaluating which part of the "non-*self*" is compatible with the survival of *self*.

By two to three years of age (Stage 2) the concept of *self* is well established, as evident in the use of the pronoun "I" and its possessive form "mine." A child is able to project *self* as an object by identifying himself in a mirror, as can be verified with the "mirror test."[3] When presented with a mirror, the child may first examine the backside and see if there is a real object behind it, but he will quickly realize that the image is his own face and body. One way to confirm self-recognition is to surreptitiously put a red spot on the child's nose ("rouge test") and see if he or she will touch his/her nose while looking at the mirror.[4] A child passing this test should also be able to pick out himself from a group picture. Great apes, dolphins and elephants also pass the mirror test, whereas dogs fail. Since the dog's world is very much determined by odor, a dog will first sniff at the mirror and then become disinterested. By contrast, a fiddler crab will fiercely fight its own image (as an enemy) in a mirror. Dolphins will look into the mirror and examine their own bodies and their mouths, very much like chimps and humans.[5] In a well-publicized case, an elephant named Happy at the Bronx Zoo in New York was marked with an "X" on the forehead where it could not see except with a mirror. The elephant repeatedly touched the spot with its trunk, while looking into the mirror, as if trying to examine it or to remove it.[6] In Stage 2 a person (or animal) is able to bring his own image out and place it among numerous other "non-self" objects, as if he had "stepped out of his body" and was looking back at himself.

By four to six years (Stage 3), there is further change in perspective, i.e., from out of one's own body into the mind of other persons. This is demonstrated by the "theory-of-mind" test.[7,8] Theory-of-mind refers to the ability of an agent to infer the mental state of another. As Premack and Woodruff put it: "An individual has a theory of mind if he imputes mental states to himself and others. A system of inferences of this kind is properly viewed as theory because such states are not directly observable, and the system can be used to make predictions about the behavior of others." Theory-of-mind was first tested on a chimpanzee using videotapes and photographs; the animal passed the test. Normal children passed but the mentally retarded failed. A typical protocol is illustrated by the "Sally-Anne Test" (Fig. 8.1).[9] Sally (the girl in white, on the left)

Fig. 8.1. "Theory-of-mind" test as illustrated by the Sally-Anne situation (see text for detail). Children beyond 6 years of age pass this test, as they believe that Sally will look for the ball in the bowl, since this is where she put it. The test subjects recognize other people's mind or mental reality. [See Note 9.]

has a ball, which she places in her bowl. After she leaves, Anne (the girl in black, on the right) surreptitiously moves it to her box. When Sally returns, where will she look for the ball? Children below four years of age answer that Sally will look for it in Anne's box, since this is where the ball is; but older children answer that Sally will look for it in her own bowl, because she mistakenly believes that the ball is still there. The ability of a test subject to dwell in other people's mental states incurs thinking about mind, or thinking about thinking. When applied to one's own *self*, it is recursive or reflective thinking, in other words, introspection.

8.2 Some Properties of Consciousness

In 1892 William James defined consciousness as a state of the mind that has four characteristics: (1) it is totally private; (2) it is constantly changing; (3) each momentary state of consciousness is continuous with the preceding and the following ones; (4) the content is focused and selective for each person. This is known as the "stream of consciousness" because it is analogous to flowing water.[10] I shall briefly review these points, with new insights added whenever appropriate.

First, consciousness cannot be observed directly. It can only be inferred from the behavior of an animal, from the workings of the brain, or, in the case of humans, from testimony. In dealing with consciousness, certain empathic influence from the observer cannot be avoided. For this reason, psychology can never be as impartial a science as physics and chemistry.

Next, consciousness has content, which is constantly changing. At the most basic level, the content is made up of sense data, which organize to form recognizable structures of the world. During our waking hours, some sort of image always stays in our mind. When we open our eyes, the image is usually that of the immediate outside world. But once the eyes are closed, some image lingers on, which may or may not correspond to what we just perceived — it could be an image of the past or one we imagine.

One of the properties of consciousness is its intensity, or focus of attention. A global but diffuse consciousness can be likened to a room illuminated by a light bulb, whereas intensified consciousness acts like a searchlight, focusing on one spot or scanning left to right. The neurophysiological basis of focusing is the selective facilitation of sensory input on one particular channel, while simultaneously inhibiting all others.[11]

Consciousness is energy consuming.[12] Therefore, it is economical for an organism not to pay attention to all details of the environment. Habits are formed to convert useful action into automatism, and *habituation* is developed to ignore repetitious innocuous stimuli, both being energy conserving. Consciousness can sometimes be compared to a landscape partially under water. What we are aware of is the part above water, but what is below can become visible by lowering the water level. A case of heightened consciousness happens when your name is suddenly announced in the public address system, or when you unexpectedly spot your loved one in a crowd.

It is interesting that only 5% of our daily activities are in the conscious realm; the rest, including such semi-conscious vital functions as breathing, are carried out automatically.[13] Subliminal stimulation is one that is short enough (a few milliseconds) not to arouse awareness but long enough to affect subsequent behavior of the subject, as used in commercial advertisement.

Every object of awareness carries a graded emotional value, marked as good, bad or neutral. Without emotion, consciousness is a barren landscape; without consciousness, emotion is a blind impulse. From the evolutionary standpoint, consciousness combined with emotion provides great survival advantage.

8.3 Electrical Correlates of Consciousness

The physical correlate of consciousness was first demonstrated in 1924 when Hans Berger recorded electrical signals from the human scalp and related them to the state of wakefulness. The electrical recordings are called electro-encephalograms (EEG), commonly known as brain waves (Fig. 8.2). When a person is awake and looking out, the EEG pattern is that of a series of very rapid, desynchronized waves, corresponding to a state of alertness. With eyes closed, the EEG pattern of an awake but resting brain is that of synchronized waves of 8 to 12 hertz (cycles per second), called alpha rhythm. When a person becomes drowsy, the waves slow down with increase in amplitude, and the rhythm is less regular. As a person goes deeper and deeper into sleep, the waves further slow, the amplitude further increases, and the rhythm tends to disappear. But when a person starts to dream, the electrical activity is energized again (high frequency, low amplitude, irregular rhythm), simulating the state of alert wakefulness, hence the name "paradoxical sleep." This stage is

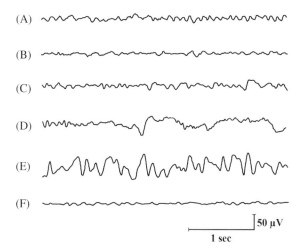

Fig. 8.2. Electro-encephalogram (EEG) of an adult human brain using electrodes placed at the back of the head. (A) Awake and relaxed (eyes closed), showing alpha rhythm. (B) Awake and alert (eyes open), showing desynchronized waves. (C to E) Stages of sleep with increasing depth. (F) Rapid-eye-movement (REM) stage of sleep, corresponding to a state of dream. [Courtesy Dr. Malcolm Yeh.]

also called "rapid-eye-movement" sleep or REM sleep, as it is accompanied by rapid oscillation of the eyeballs. REM sleep and dreaming go hand in hand. Although it can be considered a state of consciousness, the content of a dream is unrelated to the immediate surroundings, and it usually cannot be recalled unless the person is awakened at the time of REM sleep. The physiologic significance of REM sleep is not clear. We know it is essential to life, since experimental animals subjected to prolonged deprivation of REM sleep do not survive. There is evidence that REM sleep enhances memory and learning.[14]

Prior to the advent of EEG, electrical activities were detected only in isolated nerves. EEG provides a direct correlation of consciousness and physical activities inside a living brain. Since then, electrical circuitry has become the most accepted marker of consciousness. The main shortcoming of classical EEG is that the signals are weak because of the intervening skull. Nevertheless, the rhythm and variation in amplitude of the waves are very useful, as they reflect the functions of the deep brain structures, in particular the thalamus. EEG is sensitive in detecting seizure activity in the brain, even when a person is not actively in convulsion. Recent progress consists of direct recording on the exposed brain. The method becomes even more powerful when multiple electrodes are simultaneously applied. The limitation to this approach is that the procedure is invasive and needs justification when carried out on people. For this reason, direct recording on the human brain is confined to patients undergoing brain surgery.

A variant of EEG is magneto-encephalogram (MEG), which records variations in magnetic properties; it is superior for detecting deep brain activities but has less resolution power. A combination of EEG, MEG and functional MRI (magnetic resonance imaging) provides more information than any one alone.

8.4 Neurology of Consciousness

The neurophysiologic mechanisms of consciousness can be divided into two categories: those for specific sensations and perceptions (discrete mode); and those for general arousal (diffuse mode). See Table 8.2.

Table 8.2. Brain Structures Relevant to Consciousness

A. Discrete Consciousness (content relevant) —

Inputs from spinal tracts and cranial nerves project to discrete brain sensory areas such as the somatosensory, visual, and auditory cortices, following a relay in the thalamus.

B. Diffuse Consciousness (content irrelevant; general alertness) —

There are four components:

(1) Brain stem reticular activating system: Sends short fibers to thalamus which then projects to diverged cerebral cortical areas.
(2) Locus ceruleus (in dorsal pons): Sends noradrenergic-producing fibers to diverged areas including cerebral cortex, cerebellar cortex, olfactory bulb (in rats), thalamus, and hypothalamus.
(3) Dorsal raphe nuclei (spread out along brainstem): Sends serotonin-producing fibers to thalamus, cerebral cortex, cerebellar cortex, and neostriatum (caudate and putamen).
(4) Lateral hypothalamus: Sends orexin-producing fibers to posterior hypothalamus, thalamus, locus ceruleus, dorsal raphe nuclei, and the frontal lobe.

8.4.1 *Discrete mode*

Discrete sensations from the outside world start as stimuli striking the sense organs: retina of the eye; auditory hair cells of the inner ear; olfactory cells of the nose; taste buds of the tongue; skin receptors for touch, pressure, and vibration; cutaneous nerve endings for pain and temperature; and vestibular cells for movement and gravitational sense. From the sense organs the messages are transduced into nerve impulses that travel to the specific sensory areas in the cerebral cortex, by way of a relay station in the thalamus.[15] Neurophysiologic correlates of discrete sensations can be detected by *sensory evoked potentials* recorded from the respective areas in the brain.

Cortical areas for discrete sensations include: visual cortex in the occipital lobe, auditory cortex in the temporal lobe, and somatosensory cortex (for touch, pressure, pain and temperature) in the parietal lobe. (For locations of these structures refer to Fig. 7.10 in *Chapter 7*). However, perception is an active neural construct that involves more than activation of specific sensory areas. Take hearing for example. Electrical signals reaching the auditory cortex in the temporal lobe are projected to the parietal and frontal cortices, from where feedback signals are sent down to the temporal cortex. The reverberation of neural signals distinguishes perception from simple sensation, suggesting that conscious awareness of the environment requires recurrent processing of the incoming sense data in the higher-order association areas.[16] Studies on visual perception show that a human subject needs 300 milliseconds of stimulation for the process to reach the conscious state (to become visible). Concomitant electrical recording correlates the conscious state to a widespread, sustained activation of the fronto-parieto-temporal network, involving multiple recurrent cortical loops, without which only the occipital area (primary visual cortex) is stimulated to a subliminal degree.[17] Further evidence supports the top-down modulation of visual sensory processing, when it was shown that neurons from the cingulate region of the frontal cortex send fibers to the primary visual cortex to modulate discrimination of visual input. This mechanism, called "selective attention," enables the brain to focus on a particular item out of thousands of objects in the visual scene.[18]

Dehaene and Changeux proposed that perception arises as a construct of incoming sensory data based on internal models derived from previous experiences. In the process the brain matches the new with the old and assembles the former into the latter. It is a way of making sense of the present by comparing with the past. This theory, called the "global neuronal workspace model," posits that conscious awareness results from a dialogue between the sensory, motor, attention, memory, and value areas to form a higher-level, unified space where information is shared and broadcasted back to lower-level processors. The dialogue is made possible by pyramidal cortical neurons, which send long-distance cortico-cortical connections among the primary sensory areas and the association areas of the brain.[19]

Edelman stated that the brain is too complex to work logically by a computer-like algorithm. He suggested that it works by selecting (through parallel and reentrance dynamics) the relevant out of randomly formed possible representations. He pointed out that the thalamo-cortical fibers that project in a reciprocal manner fit this recursive modality, and therefore are important for conscious experience.[20] Crick, on the other hand, proposed that the synchronized firing (at about 40 hertz) between the thalamus and the cortex could be correlated with visual consciousness.[21]

As to the awareness of the bodily *self*, much can be learned from the phenomenon of *phantom limb*. The phantom limb experience is the perception of a body part when it is no longer there. The phenomenon occurs in at least 90% of amputees, often accompanied by pain, which may persist for years. Using neuro-magnetic imaging techniques, there is a correlation between the severity of phantom-limb pain and the degree of reorganization in the somatosensory cortex. The phenomenon provides additional credence to the theory that the perception of body image, like perception of the external world, is also a product of systems construction in the brain.[22]

8.4.2 Diffuse mode

Distinct from discrete sensations and perceptions, non-specific alertness is generated by an activating system in the brain stem and thalamus (Fig. 8.3; Table 8.2). The brain stem *reticular formation* is an assembly of diffuse, interconnected neuronal processes forming an ascending (toward the brain) network. These short fibers are collaterals of the long sensory neurons from the spinal cord and the brain stem. Reticular formation extends upwards to the thalamus, from where fibers are projected to the amygdala and hippocampus, and diffusely over the entire cerebral cortex without focusing on any specific sensory area. Electrical stimulation along the reticular formation promotes general wakefulness of an animal, hence the name *"reticular activating system"* (RAS). Destruction of RAS leads to coma.

Among the short, diffuse network of reticular interneurons are a few long projection fibers in the brain stem that are also involved in general arousal and sleep-wake cycles, the most well studied of which

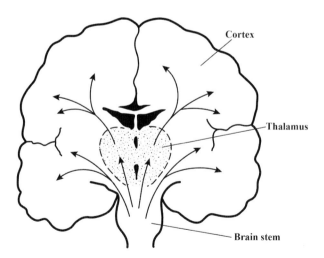

Fig. 8.3. Schematic drawing of the human brain in coronal section (vertical cross-section), showing the *reticular activating system* conveying arousal from the brain stem to the cerebral hemispheres through a relay in the thalamus. Note that some fibers bypass the thalamus.

are the *locus ceruleus* (noradrenergic system) in the dorsal *pons*, and the dorsal *raphe* nuclei (serotonin system) along the midline of the entire brain stem (Table 8.2). Both are associated with behavioral vigilance.

The lateral hypothalamus is involved in the sleep-wake cycle. Its destruction results in somnolence. This is due to a group of neurons in this region that produces orexin, a small peptide capable of arousal. Fibers from these neurons project to diverse places of the brain.[23,24]

In the aging human brain where the acetylcholine (ACh) activity is low, pharmacologic stimulation of nicotinic acetylcholine receptor (nAChR) is known to restore alertness and thus improve cognitive performance. Mice with nAChR deleted genetically have impaired attention, and the selective re-expression of the receptor in the brain corrects the deficit.[25]

8.5 Impairment of Consciousness

8.5.1 *Transient*

Absent-mindedness (lack of focused attention) is common in normal persons. Syncope (fainting) is a transient, benign condition resulting from sudden drop of blood flow to the brain. It can be brought about by postural change or by strong emotional states. Recovery is expected. EEG during syncope is normal. Drowsiness, stupor, and coma are conditions of decreasing alertness due to drug intake, metabolic disorder, brain disease, or head trauma. EEG tracing usually simulates that of slow-wave sleep. A person can either recover or die from coma within a few days, or slip into one of the following permanent conditions.

8.5.2 *Prolonged*

Minimally conscious state is a condition in which a person is awake but has only minimal awareness of the environment; interaction with people is usually absent. EEG tracings may simulate that of drowsiness or sleep. In the *persistent vegetative state* the patient has sleep/awake cycles but is totally unaware of the environment. Some EEG activity is retained, as is the cardio-respiratory function. Reflexes are present but there is no purposeful movement. The eyes may open spontaneously, though the patient shows no intention to communicate. With tube feeding, a "vegetative" patient may survive for years. (The distinction between minimally conscious state and persistent vegetative state is not clear-cut and can be a source of controversy.) *Brain death* is a condition in which the EEG is flat and the patient cannot be aroused. In brain death cranial nerve reflexes are absent but spinal nerve reflexes may be retained. Cardiovascular function and general metabolism are retained whereas respiration has to be artificially maintained. The loss of consciousness in brain death is irreversible as the neurons in the brain undergo necrosis (cell death). *Locked-in syndrome* is a condition in which the patient is conscious but unable to communicate except for minimal muscle twitching such as eye blinking. It can easily be confused with loss of consciousness.

8.5.3 *Recurrent and episodic*

Epilepsy is a group of neurological disorders typically characterized by recurrent, sudden but transient loss of consciousness (with or without muscular jerking), accompanied by an EEG picture of bursts of excessive electrical activities. One type of epilepsy, called *temporal lobe epilepsy* (the pathology being in the temporal lobe), manifests a state of altered consciousness, during which time the patient may engage in simple activities of which he subsequently has no recollection. For example, during the epileptic attack (seizure), the patient may pace aimlessly inside a room without bumping into any of the furniture. He may open his eyes but engages in no purposeful act, and does not respond to other people in the room. Because of bizarre behavior, temporal lobe epilepsy is also known as psychomotor seizure. *Sleepwalking* is a sleep disorder in which the patient suddenly gets up and wanders in the room for a few minutes while still in sleep. The patient looks confused and will have no recall of the incident. *Narcolepsy* is a pathologic condition in which a person is afflicted with repeated, episodic attacks of irresistible sleepiness.

During the attack, EEG shows the REM pattern and the patient falls into a dream state, which may intermingle with wakefulness. It is an autoimmune disease in which the orexin-producing cells (responsible for arousal) in the hypothalamus are destroyed. The disease has been demonstrated in animal models.

Notes and References

1. The subjective, qualitative aspects of sense data are called *qualia* in the philosophy of mind.
2. The most fundamental exteroception is touch, which is intuitively used by most people as a test for reality of an object, as happens when a visual image is in doubt.
3. Gallup GG Jr. (1970) Chimpanzees: Self recognition. *Science* **167**: 86–87.
4. Amsterdam B. (1972) Mirror self-image reactions before age two. *Develop Psychobiol* **5**: 297–305.
5. Reiss D. (2011) *Dolphins in the Mirror.* Houghton Mifflin Harcourt.
6. Plotnik JM, de Waal FBM, Reiss D. (2006) Self-recognition in an Asian elephant. *Proc Natl Acad Sci USA* **103**: 17053–17057.
7. Premack DG, Woodruff G. (1978) Does the chimpanzee have a theory of mind? *Behavioral and Brain Sci* **1**: 515–526.
8. Perner WH Jr. (1983) Beliefs about beliefs: Representation and constraining function of wrong beliefs in young children's understanding of deception. *Cognition* **13**: 103–128.
9. Baron-Cohen S, Leslie AM, Frith U. (1985) Does the autistic child have a "theory of mind"? *Cognition* **21**: 37–46.
10. James W. (1890) *The Principles of Psychology.* Henry Holt; republished by Dover, New York, 1950.
11. Livingston RB. (1959) in *Handbook of Physiol* Sect. 1: Neurophysiology; Field J, Magoun HW, Hail VE. eds. Am Physiol Soc Wash., D.C, Vol. 1, p. 741.
12. It is estimated that 25% of an adult person's energy, and 50% of a newborn's, is spent in the brain.
13. Baumeister RF, Sommer KL. (1997) Consciousness, free choice, and automaticity. In: Wyer RS Jr. ed. *Advan Soc Cognition* Vol. 10. Erlbaum, Mahwah.

14. Boyce R, Glasgow SD, Williams S, Adamantidis A. (2016) Causal evidence for the role of REM sleep theta rhythm in contextual memory consolidation. *Science* **352:** 812–816.
15. Most of the information about the outside world passes through the thalamus before reaching the neocortex. Exceptions to the thalamic relay include smell fibers and some taste fibers.
16. Boly M, Garrido MI, Gosseries O, *et al.* (2011) Preserved feedforward but impaired top-down processes in the vegetative state. *Science* **332:** 858–861.
17. Del Cul A, Baillet S, Dehaene S. (2007) Brain dynamics underlying the nonlinear threshold for access to consciousness. *PLoS Biol* **5:** e260.
18. Zhang S, Xu M, Kamigaki T, *et al.* (2014) Long-range and local circuits for top-down modulation of visual cortex processing. *Science* **345:** 660–665.
19. Dehaene S, Changeux JP. (2011) Experimental and theoretical approaches to conscious processing. *Neuron* **70:** 200–227.
20. Edelman GM. (2003) Naturalizing consciousness: A theoretical framework. *Proc Natl Acad Sci* **100:** 5520–5524.
21. Crick F. (1994) *The Astonishing Hypothesis*, Simon & Schuster, New York.
22. Flor H, Elbert T, Knecht S, *et al.* (1995) Phantom-limb pain as a perceptual correlate of cortical reorganization following arm amputation. *Nature* **375:** 482–484.
23. Nerve fibers carrying orexin from the lateral hypothalamus project to the following areas: *locus ceruleus*, the dorsal *raphe* nuclei, the intralaminar and midline thalamic nuclei, the forebrain cholinergic neurons, as well as the posterior hypothalamus. See: Sutcliffe JG, de Lecea L. (2006) Hypocretins/Orexins in brain function. In: Lim R. ed. *Handbook of Neurochemistry and Molecular Neurobiology*, 3rd ed. Neuroactive Proteins and Peptides, Springer-Verlag, Berlin, pp. 501–522.
24. In addition to causing alertness, orexin also stimulates appetite.
25. Guillem K, Bloem B, Poorthuis RB, *et al.* (2011) Nicotinic acetylcholine receptor $\beta 2$ subunits in the medial prefrontal cortex control attention. *Science* **333:** 888–891.

Chapter 9: *Self* and Emotion

"Men are rather reasoning than reasonable animals, for the most part governed by the impulse of passion.

— Alexander Hamilton, 1802

Overview: *Emotion (or affect) is the bodily reaction to a stimulus when such a stimulus is of relevance to the preservation of self. Emotion is the driving force for action. The desirability of a stimulus can be placed on a value spectrum representing a continuum from the most desirable to the least (the most aversive). Emotion is most intense when a stimulus is at either end of the spectrum, whereas in the middle (neither good nor bad) it is neutral and does not elicit a response.*

The physiology of emotion can be divided into a peripheral and a central component. The peripheral component (emotional outburst) is the outcome of sympathetic discharge. It is outwardly observable and internally felt by the agent. The central component, comprising the various subcortical emotional centers and part of the cerebral cortex, serves as the evaluator as to whether a sympathetic discharge should be mounted. The central component also provides a sustained background feeling (or mood) when the emotion is not outwardly expressed. The cerebral cortex is in constant negotiation with the subcortical emotional centers. When the former has the upper hand, we see reason suppressing primitive impulse; when the latter dominates, we see emotion undermining rational thinking.

Art is an extracorporeal projection of emotion. Once created, art assumes an independent existence and continues to influence the observers even in the absence of its creator.

9.1 We are All Captives of Passion

The following was taken verbatim from Bill Bryson's recent book: "If your two parents hadn't bonded just when they did — possibly to the second, possibly to the nanosecond — you wouldn't be here. And if their parents hadn't bonded in a precisely timely manner, you wouldn't be here either. And if their parents hadn't done likewise, and their parents before them, and so on, obviously and indefinitely, you wouldn't be here."[1]

Bill was being polite and subtle. If the word "bonded" were replaced by "copulated," it could have been applicable to almost all living things, down to the bacteria *E. coli* that do their thing by the scientific euphemism called "conjugation." That is how we got here, and that is how endless future generations will continue to be here. I, for one, would not be around if my countless ancestors had not lusted for each other.

We humans are slaves of desire, as are butterflies and scorpions, the only difference being our ability to convert instant to deferred gratification. Over eons, evolution gleans the instincts that preserve *self* and the species and imbues them with strong motivation, frequently irresistible: why steaks are so tasty, why boys chase girls, why babies are so cute and cuddly, why salmon swim upstream hundreds of miles to the point of exhaustion, only to find a place to spawn and die. History is littered with episodes of people taking great risks, often at the peril of their wealth, power, reputation, and even life, only to release a moment of rage or satisfy an irresistible desire. Remember how Helen ignited the Trojan War, how Henry VIII broke with the Vatican, how King Edward VIII abdicated his throne, and how many high level government officials met their downfall by engaging in socially unacceptable affairs?

Emotion is the driving force to action for the benefit of the *self*. It is a strong undercurrent motivating all behaviors, however rational the action may appear. Emotion is to consciousness as color is to shape. Consciousness without emotion is an aimless, drifting experience; emotion without consciousness is an impulse devoid of content. Emotion is deeply ingrained in the animal world. Thanks to this faculty, we animals can procure food, avoid danger, and beget progeny more efficiently than a cabbage.

9.2 What is Emotion?

First, let me clarify a few terms I use in this chapter. The word "emotion" when used without qualification denotes the general *affective* aspect of the mind as opposed to the cognitive function such as knowing and rational thinking. I use the word in its broadest sense, to include such things as joy, sorrow, love, hatred, fear, jealousy, and the like. I distinguish two aspects of emotion: one is the peripheral and outwardly response, which I shall call "emotional expression" or, in the extreme case, "emotional outburst;" the other is the inner experience with or without an outward expression, which I shall call "feeling, or mood." Emotional expression includes facial features showing anger, horror, surprise, pleasure, or happiness, as well as involuntary bodily reactions such as palpitations, rapid respiration, muscle tension, or intestinal spasm. "Feeling" is the subtle aspect of emotional experience without overt bodily components. It is a high-order mental function derived from the interaction of cerebral cortex and the subcortical emotional centers. "Mood" is a prolonged feeling of a diffuse nature that serves as a tonal background.[2]

9.3 Where is the "Seat" of Emotion?

Common sense tells us that the heart is where we feel the emotion. Everybody understands when your heart is "broken," or so-and-so is "kind-hearted," or when you thank someone from the "bottom of your heart." When excited, angered, frightened, or deeply in love, you feel it in your heart. When Cupid (meaning *desire* in Latin) shoots his arrow at a young man and woman, it is the heart that he aims at. That the heart is the seat of emotion is deeply engrained in all folk culture. Four thousand years ago, when the Egyptians mummified their dead, the heart was carefully preserved in a jar and placed next to the body while the brain, thought to be junk, was simply discarded.

The focus of attention was later shifted to the brain when it was realized that the brain is responsible for all mental functions. In the seventeenth century, Descartes maintained the brain as the organ of the mind but put it under the control of the soul. In this dualistic view, the physical body, which includes the brain and the heart, is nothing but a machine, driven into motion by a spirit sitting comfortably in the center of the brain, in a small structure called the pineal gland. The pineal is unique among brain structures in that it occurs singly and is in the midline on the top of the midbrain, protected on both sides by the two large cerebral hemispheres.[3] Descartes regarded the pineal as the "driver's seat," a place where mind (soul) and matter (body) interact and negotiate. Emotion, like cognitive function, in Descartes' view, is the business of the soul.

As human biology and psychology progressed, scientists tried to work out the mechanism of mind without having to resort to a nonphysical soul. By the late nineteenth century, the cognitive function of the cerebral cortex was known, and it was widely believed that the emotions are the results of cerebral activity. William James, a noted American psychologist, stunned the world by going against the prevailing tide. He proclaimed that emotion is nothing but the sensory experience of bodily changes (heartbeat, blood pressure, respiration rate, muscular tension, sweating, crying, etc.) in response to exciting stimuli, thus placing the site of emotion back in the periphery, away from the brain. In James' oft-quoted statement, "If we fancy some strong emotion, and then try to abstract from our consciousness of it all the feelings of its bodily symptoms, we find we have nothing left behind, no 'mind-stuff' out of which the emotion can be constituted, and that a cold and neutral state of intellectual perception is all that remains."[4]

James made a seminal contribution to psychology by stripping off the spiritual nature of emotion and planting it on the solid ground of matter — the body. Simultaneously with Carl Lange of Denmark, James suggested that an emotion-arousing stimulus is first received by the sensory cortex (without emotion or feeling), then channeled to the periphery where a host of physiological reactions occur, and the subsequent experiencing by the brain of these bodily changes is the emotion (known as the "James-Lange Theory of Emotion"). However, James went overboard by saying, "We feel sorry because we cry, angry because we strike, afraid because we tremble, and not that we cry, strike, or tremble, because we are sorry, angry, or fearful, as the case may be."[5]

From our current viewpoint, the shortcoming of James' theory lies not in the emphasis on peripheral reactions but in the absence of a central controlling mechanism. First, it lacks an evaluative function preceding or concomitant with the peripheral reactions.[6] The weighing of the survival value (or biological desirability) of the perceived stimulus is crucial to the function of emotion. Second, James' theory cannot explain how recalling a past event can elicit an emotion, in the apparent absence of an immediate outside stimulus. Third, James was ambiguous about the inner feelings of an emotional experience. He did not distinguish the perception of bodily changes from the inner feelings on a higher plane.[7]

In hindsight, we can understand why William James did not bet his money on the central nervous system. In the late nineteenth century, when James posited his hypothesis, very little was known about the interior of the brain. The cerebral cortex was demonstrated to have only two main roles: the sensory areas, which receive outside information, and the motor cortex, which controls the skeletal muscles. The rest of the brain was a "dark continent." At the time, there was no specific spot in the brain where James could securely anchor an emotional theory. James had done a great job with whatever scientific knowledge was available to him. Had he been born in our era and known as much neurophysiology as we now know, I am certain he would have presented a more complete theory and would have toned down such strong statements as "we feel sorry because we cry, anger because we strike…"

It is indisputable that no emotional experience is complete without sensing the physiological changes in the body, but these changes do not comprise the totality of an emotional process. Between the perception of an external object and the elicitation of the physiological changes, there has to be an intervening brain mechanism, of whose workings the agent may or may not be aware, that evaluates the survival value of the object perceived with respect to *self*, the outcome of which evaluation results in the intensity of the physiological reactions; the intensity being proportional to the degree of positivity or negativity of the valuation.

9.4 Autonomic Nervous System and the Hypothalamus

Some important events took place in the ensuing fifty years following the formulation of the James-Lange theory. One was the discovery of the *autonomic* nervous system and its relation to the *hypothalamus*. The other was the discovery of the "emotional centers" in the brain.

It has been known for a long time that not all activities of the animal body are under volitional control. For example, your heart beats 75 times a minute, your breathing goes on whether you are awake or asleep, and you sweat profusely when the weather gets hot, all without your conscious effort. Such involuntary, relatively stereotyped activities are under the control of two opposing forces, together called the *autonomic nervous system* (Fig. 9.1). One part, called the *sympathetic* nervous system, is responsible for the expenditure of energy in times of heightened awareness or emergency; the other, called the *parasympathetic* nervous system, is responsible for restoration and conservation of energy in restful times. The sympathetic responses include: widening of the pupils, thirst, acceleration of the heart, rise in blood pressure, rapid shallow respiration, dilatation of the bronchioles, constriction of the arterioles (except those in skeletal muscles), sweating in the palms, increased level of blood sugar, inhibition of gastrointestinal and urinary functions, and increased skeletal muscle tone. All these changes enable the subject to escape danger, to overpower an opponent, or to procure a delicious meal. The parasympathetic responses are the opposite of the above. The autonomic nervous system not only expresses its responses on the end organs, but also reports back from the end organs to the brain via sensory fibers on a conscious level, thus producing the sensation of palpitation, sweating, abdominal cramps, etc., during stress.

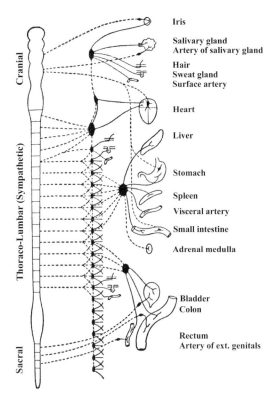

Fig. 9.1. Diagram of the autonomic nervous system. The spinal cord is the vertical elongated structure on the far left. Solid circles outside of the cord are autonomic ganglia. There are two sets of ganglia: (A) the sympathetic ganglia which receive neurons from the thoracic and lumbar regions of the cord (thoraco-lumbar outflow) and are aligned in parallel with the cord; (B) the parasympathetic ganglia which receive neurons from the head (cranial) and the tail (sacral) regions of the central nervous system, and are distributed near the end organs away from the cord (cranio-sacral outflow). The first set of ganglia contains sympathetic neurons, whereas the second contain parasympathetic neurons. From the ganglia, post-ganglionic fibers come out to enervate the end organs (far right). The general rule for neurotransmitters is: parasympathetic fibers (both pre- and post-ganglionic) and sympathetic pre-ganglionic fibers release acetylcholine (ACh) (dotted lines), whereas sympathetic post-ganglionic fibers (solid lines) release noradrenaline (norepinephrine). Both sympathetic and parasympathetic outflows receive commands from the hypothalamus in the brain through the spinal cord. The sympathetic outflow includes the adrenal medulla (a pair of endocrine glands located above the kidneys), which receives pre-ganglionic fibers from the spinal cord and releases adrenaline (epinephrine) into the bloodstream. Bear in mind that the autonomic nervous system also contains a sensory component that reports end organ responses back to the brain; it is this type of feedback sensation (interoception) that constitute the subjective aspect of an emotional expression. [See Note 51: Cannon; after Bard, Note 52; courtesy Clark Univ. Press.]

It was immediately apparent that many of the physiologic reactions, and the ensuing sensations of these reactions, so central to James' peripheral theory of emotion, were identical to the motor (efferent) and sensory (afferent) activities of the autonomic nervous system. Walter Cannon, along with his student Philip Bard, was mainly responsible for bringing the autonomic nervous system to light. They went one step further by pointing out that one essential function of the sympathetic system is to prepare the animal for a quick and violent action, the so-called "fight-or-flight" response to a threat. Cannon and Bard identified the *hypothalamus* as the brain center controlling the autonomic nervous system. They also referred to the brain for "subtler emotions" such as sustained feelings, a function that the James-Lange theory failed to account for. Thus, the Cannon-Bard theory shifted the center of emotion back to the brain.[8] The hypothalamus also mediates the secretion of cortisol, the stress hormone from the adrenal cortex (an endocrine gland at the upper pole of the kidney), through the hypothalamo-pituitary-adrenal axis, though this is a much slower process than sympathetic discharge.

9.5 Brain Centers Concerned with Emotion

In 1939, Kluver and Bucy demonstrated that bilateral removal of temporal lobes in monkeys produced dramatic behavioral alterations. Among many changes, their affect flattened, they became tame and placid when approached by humans, and their sexual appetite increased enormously, including mounting inappropriate objects and members of other species. Although the operation was crude and the lesion too extensive by today's standard, this was the first evidence that emotional centers are present inside the brain.[9]

The next moment of truth came in 1943, when reward centers were serendipitously discovered. Olds and Milder inserted electrodes into the rat brain. To their astonishment, they observed that stimulation of some areas motivated the animal to incessantly self-stimulate (by pressing a bar to allow electric current to be delivered). The animal would self-stimulate as often as 700 times per hour. The reward could be so strong that the animal would even forego food and sex. These areas were interpreted as producing pleasure, as their stimulation apparently gave the animal great satisfaction. On the other hand, there were areas (aversive centers) that the animal would avoid stimulating, and there were other areas that were motivationally neutral.[10]

Since the initial observations of Kluver and Olds, a number of emotion-related brain centers (both subcortical and cortical) have been identified. Some of these are described below. Most of these structures are included in the so-called *limbic system*, roughly equivalent to the emotional part of the brain. (For anatomical locations, refer to Figs. 9.2 and 9.3 of this chapter, and Figs. 7.10 and 7.11 of *Chapter 7*.)

9.5.1 *The amygdala*

Among the emotional centers the *amygdala* is probably the most studied (Figs. 9.2 and 9.3). Named in Latin for its almond shape, the amygdala is a large group of *nuclei* deep in the medial temporal lobe.[11] Because of its size and complex organization, the amygdala plays a pivotal role in many emotional events, the most prominent of which is *fear and anger*. In retrospect, many of the behavioral changes observed in the Kluver-Bucy experiment can be attributed to the removal of the amygdala. The amygdala receives multiple inputs and projects to the hypothalamus for autonomic emotional expression.[12] It facilitates stereotyped emotional movements such as freezing, fleeing, and biting. Synaptic changes have been detected in the amygdala following classical fear conditioning training.[13] The reciprocal connection between amygdala and the ventromedial prefrontal cortex explains the interaction between emotion and decision-making. Its input to the hippocampus allows for storage and retrieval of emotionally charged experience. Psychiatric disorders such as anxiety, depression, post-traumatic stress disorder, phobias, and even autism and schizophrenia, have been ascribed in some way to amygdala dysfunction. Interestingly, the size of amygdala has been positively correlated with the degree of aggressiveness across animal species.

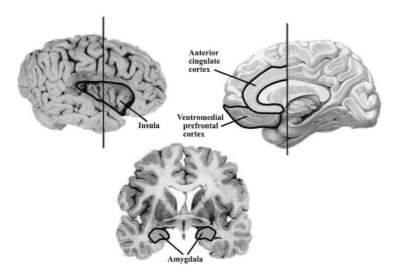

Fig. 9.2. Human brain showing some areas involved in emotion. **Top left:** Right hemisphere seen from the outside, with part of the frontal cortex carved away to expose the insula. **Top right:** Right hemisphere seen from the inside, showing the anterior cingulate cortex and the ventromedial prefrontal cortex. **Bottom:** Cross-section of both hemispheres cut along the vertical line indicated in top pictures, showing the location of the two amygdalas. Note that other deep brain structures concerned with emotion, such as nucleus accumbens, are not shown in this figure. [See Note 22; permission Cell Press.]

9.5.2 *Nucleus accumbens and ventral pallidum*

The role of the *nucleus accumbens*, also known as *ventral striatum*, located deep in the basal forebrain, as a principal "pleasure center" is well documented (Fig. 9.3). Its function has been correlated with sexual activity, food intake, and even artistic enjoyment.[14] It was reported that insertion of electrodes into the nucleus accumbens in humans alleviates severe depression, and that the nucleus accumbens is central to the mechanism of placebo effect. The major input to nucleus accumbens is the dopaminergic pathway from ventral tegmental area (VTA) in the midbrain. Injection of dopamine directly into the nucleus accumbens in the rat has been observed to increase the motivation for reward.[15] In the light of present knowledge, Olds and Milder's initial observation of

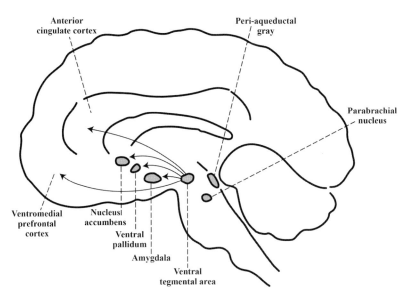

Fig. 9.3. Schematic drawing of the human brain (right hemisphere facing left) showing the approximate locations of the emotional centers. The insular cortex is out of sight in this view. Arrows show the dopamine-producing neurons from the midbrain ventral tegmental area (VTA) projecting to the various emotional centers, collectively called the "dopaminergic pathways." Within this system, those fibers extending from VTA to the nucleus accumbens, called the "mesolimbic pathway," is particularly important for motivation. A small circuit (not shown) connecting the nucleus accumbens, ventral pallidum, and ventromedial prefrontal cortex specifically generates the sense of pleasure or reward. [See Note 20, Kringelbach and Berridge.] The ventromedial prefrontal cortex also sends glutamate-carrying inhibitory fibers (not shown) to counteract the activities of the subcortical emotional centers including the nucleus accumbens and the ventral tegmental area. [See Note 53.]

self-stimulation can be explained by the activation, directly or indirectly, of nucleus accumbens-related function.

In a study conducted at Emory University, 27 teenagers were asked to listen to 120 obscure songs from unsigned artists, while their brains were scanned with functional MRI for reactions. It was subsequently observed that the activities in the nucleus accumbens were predictive of the future sales of the albums in the next three years.[16] It was shown in another study that listening to music increases activity of the nucleus accumbens.[17]

In a study regarding romantic love, male and female subjects were shown photographs of their lovers' faces. Brain imaging revealed activation of brain areas related to the nucleus accumbens.[18] Interestingly, people who experienced rejection in romantic love also showed activation of the same brain areas, suggesting that the craving/reward system is active in romantic passion, whether or not it is fulfilled.[19]

Recent investigation identified a cerebral nucleus, called the *ventral pallidum* (Fig. 9.3), that forms a chemical circuit (involving enkephalin, an endogenous opium, and anandamide, an endogenous marijuana) with nucleus accumbens and ventromedial prefrontal cortex. It was reported that the dopaminergic function of nucleus accumbens alone is responsible for *motivation*, whereas the combined action of three structures (nucleus accumbens, ventral pallidum, and ventromedial prefrontal cortex) brings in the experience of *reward*. (Note that "motivation" is synonymous with drive, craving, and wanting, while "reward" is identical with enjoyment, pleasure, fulfillment, and satisfaction.)[20]

9.5.3 *Ventral tegmental area* (VTA)

The VTA is a group of neurons in the midbrain that projects dopaminergic fibers to the nucleus accumbens, ventral pallidum, amygdala, anterior cingulate, and the ventromedial prefrontal cortex, collectively called the "dopaminergic mesolimbic system." The fibers connecting to the nucleus accumbens are particularly important for motivation behavior.

9.5.4 *Insular cortex*

The *insula* is an "island" of cortex hidden and squeezed between the temporal and fronto-parietal lobes (Fig. 9.2; see also *Chapter 10*, Fig. 10.1). Functionally, it can be considered an inward extension of the somatosensory area of the parietal lobe, with the distinction that it is confined to sensation from the internal organs (following relay in the thalamus), and thus is intimately related to the emotional life. It is here where visceral sensations are brought to the conscious level, and where an *inner image* of the current state of the body is constructed. By presenting this information to other emotional centers, including the anterior cingulate cortex, the visceral sensations are interpreted and sent to the hypothalamus before mounting an emotional discharge.[21] Using imaging studies, the insula is found to be active not only when experiencing pain or disgust, but also when seeing other persons in pain or in disgust, the latter instance suggesting a role in empathy, a socially related behavior.[22] The insula is involved in addictive behavior as well.[23]

9.5.5 *Anterior cingulate cortex*

The anterior cingulate cortex (Figs. 9.2 and 9.3; see also *Chapter 7*, Figs. 7.10A and 7.11) is intimately related to the insular cortex, receiving input fibers from the latter. It makes connections with the amygdala and the thalamus, and strongly participates in the noxious sensation of *chronic pain*. Stimulation of this area leads to stereotyped pain responses such as cardiac slowing, pupillary dilation, pilo-erection (erection of hair), and shrill vocalization, while its ablation abolishes the emotional reaction to pain.[24]

9.5.6 *Ventromedial prefrontal cortex*

The prefrontal cortex is the part of the frontal lobe excluding areas relating to the motor functions and speech; it is capable of rational thinking and executive function. The lower and midline portion of the prefrontal cortex is called the *ventromedial prefrontal* cortex, also frequently referred to as the *orbitofrontal* cortex (Figs. 9.2 and 9.3; see also *Chapter 7*, Fig. 7.11). Being a part of the rational brain that has intimate connections to the subcortical emotional centers, the ventromedial prefrontal cortex serves as a liaison between rationality and emotionality. On the one hand, its judgment function controls and dampens animal impulse; on the other, it allows temperament to penetrate into reasoning.[25,26]

The ventromedial prefrontal cortex is also involved in aesthetic experience. It is activated not only during music and visual art appreciation, but also by mathematical equations, when these equations are perceived as "beautiful" by test subjects who are mathematicians.[27]

9.5.7 *Periaqueductal gray*

The periaqueductal gray is the gray matter in the midbrain that performs a "gating" function for incoming pain sensation reaching the brain (Fig. 9.3). Its destruction leads to coma. Stimulation of this area elicits defense response such as excitement and freezing. Through its connection with the hypothalamus, it enhances copulatory behavior (lordosis) in female animals.[28]

9.5.8 *Parabrachial nucleus*

Situated in the pons, the parabrachial nucleus (Fig. 9.3) is a major channel for visceral sensation to reach the brain. It receives sensory fibers from the body after a relay in the *nucleus tractus solitarius* (NTS) in the medulla, and sends fibers to the other emotional centers including the periaqueductal gray, amygdala, hypothalamus, and thalamus. Fibers passing through the thalamus reach the anterior insular cortex, producing conscious sensation of the internal organs and the subjective feeling of emotional experience. As part of the reticular activating system it also conveys alertness. Together with the thalamus, amygdala, and hypothalamus, it also plays a role in regulating cardio-respiratory function.

9.6 Interoception and the Sensation of Emotion

Charles Sherrington in 1907 coined the word *"interoception"* to denote those sensations that refer to our own body, including the internal organs (visceral sensation), as opposed to those that refer to the outside world.[29,30] The latter, called *exteroception*, maps out the environment surrounding us through external sense organs like the eyes, ears, and skin. In contrast, interoceptive visceral sensations, which include palpitation, chest oppression, shortness of breath, stomach cramps, and intestinal spasm, are mainly outcomes of sympathetic discharge that feed back to the brain through sensory components of the autonomic nervous system.[31] Anatomically, most visceral sensations are carried in the afferent fibers of the Vagus nerve to a relay in the *nucleus tractus solitarius* (in medulla), from where fibers are sent to the parabrachial nucleus (in pons) and finally reaching the insular cortex for conscious experience. This feedback sensation from autonomic activity is what William James referred to as "emotion" a hundred years ago.

9.7 Role of Chemicals and Hormones

A number of chemicals and hormones are crucial to the mechanism of emotion in humans.[32] Dopamine is the major neurotransmitter that activates the nucleus accumbens and other areas of the mesolimbic pathway, the motivation/reward center of the brain. When infused into the brain it promotes drive and reduces sadness. Oxytocin, a hormone secreted by the posterior pituitary, enhances trust and attachment among persons, including mother-and-child, sexual partners, and friends. Vasopressin, another hormone from the posterior pituitary, increases sexual arousal and attraction and decreases anxiety. Lack of serotonin is correlated with depression, anxiety, and obsessive thinking. Adrenaline and cortisol in the blood promote alertness and decrease pain sensitivity. Testosterone stimulates aggressive behavior and increases libido in both sexes. In rats, repeated social stress promotes the release of adrenal glucocorticoids (such as cortisol), and induces social aversion and anxiety through action on the nucleus accumbens and frontal cortex.[33] The number of mood-changing and psychotropic drugs, along with their endogenous counterparts (endorphins, enkephalins, the endocannabinoids, etc.), is too numerous to enumerate here. (The role of hormones in social behavior is further elaborated in *Chapter 12: The Expanded Self: Society as Self*.)

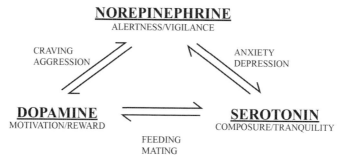

Fig. 9.4. Diagram showing very rough correlation of neurotransmitters with emotion and mood. The proper function of the transmitters is essential for a balanced emotional life.

The enzyme monoamine oxidase A (MAO-A), which breaks down the monoamine neurotransmitters norepinephrine, serotonin, and dopamine, has been negatively correlated with aggressive behavior across species. In healthy men, the lower the MAO-A activity in cortical and subcortical brain regions, the higher the trait aggression and predicted antisocial behavior.[34] In a 30-year longitudinal study, people with the low-activity variant of MAO-A genotype are more likely to develop offending conduct and hostility.[35]

Pheromones, the volatile chemical signals between animals perceived through a special organ in the nose (vomero-nasal organ), activate the subcortical center amygdala. In lower animals they are potent stimulants of instinctive responses (especially the emotion of lust) at a subconscious level. These socially relevant chemical signals are found in discharges like tears, sweat, vaginal secretions, and urine.

The effect of neurotransmitters on emotion and mood is depicted in Figure 9.4.

9.8 Appraisal Precedes Emotion

Early in the twentieth century, Cannon and Bard showed that animals could be induced to express anger and defense reactions by nonthreatening stimuli if their neocortex was removed. This phenomenon, called "sham rage," demonstrates that outward emotional expression is possible without a biological purpose.[8] Humans, also, can be manipulated into a state of euphoria, anger, or amusement in the absence of a good reason.[36] However, for an emotional response to be biologically adaptive and useful, a high level evaluator of the stimulus is required.

Normally, every stimulus carries a survival value. Most stimuli are neutral and therefore ignored, but some are either good or bad and should be reacted to accordingly. It is as if a value spectrum goes from the worst at one end to the best at the other. Emotion provides a driving force to act when the value approaches either end. The presence of an appraiser avoids the unnecessary waste of energy when a response is not called for.

Figure 9.5 illustrates my point. Picture yourself as a wild animal in the African savanna, where one either eats others or is eaten by them. When a zebra encounters a lion (predator), it has to turn away and run fast to save its life (an escape behavior). But when a lion sees a zebra (prey), it will also run, this time in the direction of the object (a chase behavior). Both are moving at top speed, with respiration and heartbeat working at full throttle, and adrenalin gushing through the blood stream.

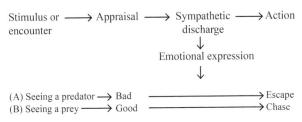

Fig. 9.5. Appraisal function in an emotional situation. An appraiser evaluates the adaptive value of a stimulus and decides whether an animal should respond positively, negatively, or not at all, and with how much intensity. Note that the appraiser not only determines the forcefulness but also the direction of the action. Emotional expression resulting from sympathetic discharge is quite similar even for opposite situations.

Both have heightened sympathetic activities, yet one is fleeing for its life and the other lusting for a meal. The motivations are diametrically opposite, yet their outward emotional expressions are for the most part similar. For the zebra, in particular, the direction of flight is a matter of life or death, and has to be decided instantaneously. Thus, emotion alone without a clear-cut goal is biologically counterproductive.

It is hard to pinpoint any particular brain structure as the site of appraisal. The good-versus-bad valuation is likely the outcome of integrative function of the cerebral cortex and the multiple subcortical emotional centers, most likely involving the hippocampus for the participation of memory and learning. It can also be genetically programmed, such as fear of snakes in many animal species or the fear of flying predators in ducklings (see *Chapter 7*, Fig. 7.13). The task may be carried out at the fully conscious, minimally conscious, or subconscious level. When valuation turns subconscious, the behavior becomes a habit or phobia. For an unexpected stimulus, the default instinctive appraisal seems to be that of aversion, as happens when a crawling insect suddenly appears on your back.

9.9 An Updated Theory of Emotion

Figure 9.6 summarizes a reasonable mechanism of emotion. Emotion in a broad sense can be divided into two components: (1) A peripheral component (outbursts) consisting of autonomic sympathetic output from the hypothalamus. The resulting bodily responses give rise to the observable outward manifestations and the inner visceral sensations (*interoception*). (2) A central component consisting of the cerebral cortex and the emotional centers. When an external stimulus (*exteroception*) arrives at the neocortex on a conscious level, the message is passed on to this central component for appraisal. If the value is either good or bad, it is conveyed to the hypothalamus to trigger a sympathetic discharge. If the emotional expression is consciously or unconsciously suppressed, it may elicit only a subtle mood change

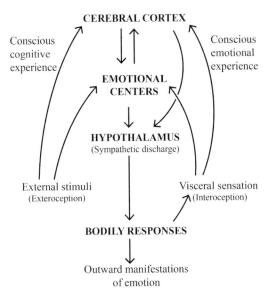

Fig. 9.6. Mechanism of emotion, showing the interactions between the cerebral cortex (rational thinking), the emotional centers (impulsive drive), the hypothalamic sympathetic discharge (emotional expression), and the interoceptive feedback from the visceral organs (emotional experience). Most sensations, external and internal, arrive at the cortex at the conscious level, but some go directly to the deep emotional centers, such as the amygdala, eliciting a response without awareness. Note that some emotional centers are in the cerebral cortex.

without an outward sign. The stimulus can also be endogenous, such as physical illness (e.g., indigestion) or brain chemistry imbalance (e.g., depression and anxiety). Furthermore, recall of an earlier experience, with input from the hippocampus, can also activate a feeling, a mood, or an outright expression. In some instances, emotion can be evoked with only subliminal awareness, bypassing the higher cortical functions.[37]

The two-way traffic between the cerebral cortex and the emotional centers permits a mutual influence between rational thinking and emotion. Thus, not only can reasoning reign over animal impulse, as in deferred gratification which we so often see at the human level, but primitive urge can also undermine rational decision. David Hume emphasized that reason alone cannot motivate human behavior, but rather humans are the "slave of passions."[38] Wundt in 1907 asserted the primacy of affective reactions over rational thinking, stating that rational thinking often follows an action in order to justify it, sometimes even to falsely justify it.[39] In an article carrying the revealing title *Preferences Need no Inferences*, Zayonc in 1980 expounded on the fact that people frequently make important life choices without knowing why they do it. He stated: "People do not get married or divorced, commit murder or suicide, or lay down their lives for freedom upon a detailed cognitive analysis of the pros and cons of their actions"[40] Even in the judicial system, decisions can be tainted by non-rational factors. The US Supreme Court Justice Oliver Wendell Holmes once remarked that "the life of the law has not been logic; it has been experience." It was argued that the rational application of legal reasons does not sufficiently explain judicial decisions, and that psychological and other factors influence rulings as well. A recent paper published in the *Proceedings of the US National Academy of Sciences* reported that judges are more lenient to the offenders when they hear the case immediately after taking a meal than when they are hungry and tired.[41]

9.10 Emotional Interpretation of Pain and Pleasure

Pain is unique among sensory modalities in that it has strong existential overtone.

As the saying goes, "if it hurts, it's me." A limb loses its intimacy if it no longer feels pain (as in stroke), and people in excruciating pain often prefer to die rather than live. Ordinarily, pain delineates our body image, and helps distinguish physical *self* from non-*self*. There are two types of pain, the discriminative and the affective. It is the second type that is inseparable from emotion. The same stimulus feels more painful if it occurs in a highly disagreeable background than one that is not. So is one that is more life-threatening. The emotional interpretation of pain takes place in the anterior cingulate cortex on the inner side of the cerebral hemisphere. Pain fibers from the internal organs pass through the thalamus and arrive at the insular cortex, and finally present to the anterior cingulate cortex to create the sensation of pain. Hypnotic suggestions to enhance or alleviate emotional pain are correlated with an increase or decrease in the activity of the anterior cingulate.[42] Watching pictures of other people suffering also activates this brain area, suggesting its role in empathy. Interestingly, the anterior cingulate is also activated in response to pleasant touch, such as caressing.

The neurologic condition called *thalamic pain syndrome* is seen in patients suffering from damage to the thalamus, resulting in relentless pain of a contralateral limb or an entire half of the body. The pain ranges from dull ache to severe, unremitting burning, is often spontaneous and is strongly emotional in character. The extreme feeling of unpleasantness contrasts with the concomitant loss of discriminatory function, such as telling a sharp from a dull point. An opposite condition is *pain asymbolia*, found in patients with damage to the insula or anterior cingulate cortex. Although these people can accurately describe the location, nature, and intensity of the pain stimulus, they lack the emotional response, and therefore they simply do not suffer.

Another interesting observation is that although pain is unpleasant, the lessening of pain is actually enjoyable, suggesting the relativity of pain. As Konrad Lorenz once quipped, there is nothing more pleasurable than a diminishing toothache.[43] In contrast, unfulfilled craving for a rewarding experience could be a tremendous suffering, as in lovesickness. In this sense, pain and pleasure can be viewed as polar extremes of a continuum. Inside the brain, the interpretation of pain and pleasure most likely involves complex, overlapped circuitries of neuronal crosstalk, including such structures as the ventromedial prefrontal cortex, amygdala, and nucleus accumbens, among other things. I should emphasize that the function of each structure should not be taken in isolation, but in the context of the overall network. In light of this, it is not surprising that a single area can be involved in opposite functions. For example, the amygdala, which usually is related to fear, is activated in some pleasurable instances, and the nucleus accumbens does not always act as a pleasure center either.[44]

9.11 When Emotion Goes Awry

Emotion loses its adaptive value if it goes to the extremes. At one end is the psychiatric condition called mania in which a person finds himself fully charged with unfocused energy. He may go several days without sleep and is compelled to do things aimlessly. He can be in sustained elation and may squander thousands of dollars in a spree, or take risks that he will later regret. At the other end is the severely depressed person who feels totally dejected and sees the world closing in on him. He may have either insomnia or excessive sleepiness or an alternation of both. He may refuse to take food and may indulge in drugs. Suicide is a common outcome. A third type of patient may display no emotion at all, as in schizophrenia. For him the outside world is totally uninteresting — neither desirable nor despicable. He displays a flat affect and everything is distant and cool. The genetic and biochemical derangements in these conditions are beyond the scope of this chapter.

In the rare affective disorder called Capgras syndrome, the patients recognize their family members, but are unable to attach any feeling to them. They know who their relatives are, but treat them as strangers in emotional terms. One patient even thought that her husband was an imposter, another person wearing the same facial features. The condition can be hereditary, but can also be seen following brain damage in the border between the temporal and parietal lobes.

9.12 Art as Extracorporeal Expression of Emotion

I have discussed emotional expression in terms of sympathetic discharge and its bodily manifestations. While this is true with all animals, humans in addition can express their emotion through objects that, if properly perceived, are capable of evoking similar feelings in others. I am talking about art. I would like to emphasize two points. First, art exerts a strong cohesive force for social bonding, which is important for the success of a group (a collective *self*) in a competitive world; this point will be elaborated in *Chapter 12: The Expanded Self: Society as Self*. Second, art has biological roots. I assume some form of aesthetic experience is present across species, the difference being in degree rather than kind. For example, highly elaborate and energy-draining courtship rituals are present in all members of the animal kingdom. Nature would not have invested in such costly behaviors if they were not essential for the continuation of species. Frogs croak, owls caw, and humans serenade under the moonlit sky, all to a common theme of affection, longing, yearning, and passion, awaiting to be consummated. Imagine how many immortal poems can retain their luster, and how many novels and operas will remain soul stirring, if their amorous theme is stripped away. It is true that art in the human world conveys much more than erotic feeling, but just the same, all forms of art — music, painting, dance, literature, and even movies — are expressions of emotion through an external medium.[45,46]

What is the neurological basis of art? Scientists were at a loss facing this problem, but evidence is accumulating rapidly that art does activate the motivation/reward system of the brain. Using functional MRI on live human brains, scientists have discovered that music enjoyment activates a network of mesolimbic reward centers including the nucleus accumbens, the ventral tegmental area, and the ventromedial prefrontal cortex, along with dopamine release.[47] Like music, visual art activates the same brain areas, even in the same person, confirming a common mechanism for aesthetic enjoyment.[48,49]

Nonetheless, it remains to be explained why some sensory perceptions (in case of art) lead to activation of the reward centers whereas others (non-art) do not. For example, we do not know how certain combinations of harmony, rhythm, timbre and melody sparks the enjoyment of a piece of music; or how certain placement of colors, shapes, and lines triggers the pleasure of viewing a painting.

Table 9.1. Echelon of Emotions

(In increasing order of lasting and overwhelming effects)

First tier — Represents instinct-driven primitive urge; involves the brain stem, the limbic system, and the emotion centers: (A) For survival of individual: Hunger, thirst, appetite, gusto, impulse, defensiveness, fear, flight, panic, rage, aggressiveness, menace, surprise, anger. (B) For survival of species: Sexual attraction and urge, lust, bonding, mating impulse, brooding and protection of the young.

Second tier — Represents interplay of higher brain functions and primitive urge; involves the prefrontal cortex, the associative cortex (parietal area), and the emotion centers: Family and filial love, obedience, admiration, hatred, aversion, grudge, joy, sadness, anxiety, jealousy, guilt, shame, regret, disdain, disgust, despondency, animosity, thankfulness, friendship, comradeship, courage.

Third tier — Represents deep-rooted, sustained mood and sentiment; involves a cortex-dominated global brain function: Depression, melancholy, worthlessness, resignation, agony, blissfulness, compassion, sympathy, altruism, self-confidence, self-esteem, assertiveness, fulfillment, elation, euphoria, ecstasy, composure, tranquility, equanimity, serenity, adoration, devotion, patriotism, religiosity.

Note: The first-tier emotions are those outwardly observable, many of which such as craving for food and sex can be satiated. Emotions in the third-tier are deep feelings that linger on for a long time. The relation to survival and procreation is most obvious in the first, but is only subtly hinted in the third. In reference to arts, those capable of arousing deep feelings (as in a tragedy) are more touching and long lasting than those that merely induce superficial funniness (as in a situation comedy). The third tier emotions include what some would call "life-changing" and what others would consider "spiritual." Because of language imprecision, the ranking is only relative; overlaps are unavoidable.

What is particularly puzzling is how the brain resolves conflicting feelings when experiencing art. For example, it has been recognized since the Greek philosophers' time that people enjoy tragedies despite their sad endings. It seems a contradiction that a person can be happy (enjoying the drama) and sad (sharing the misfortune of the protagonists) at the same time. The issue can be resolved if we consider emotions as occurring in layers (see Table 9.1). Those with global and long-lasting effects override those that are fleeting. At the lowest level are the emotions directly derived from instinct and sense organs. A sophisticated audience, on the other hand, seeks higher-level gratifications — those that can endure the passage of time, sometimes even of a "life-changing" type. Thus, the empathic mood of Picasso's Blue Period outweighs the miserable plight of the subjects, and the invigorating beats of Stravinsky's *Rite of Spring* override the cacophony of sounds.

There is another aspect of art appreciation that is in line with neuropsychological principles, and that is: the brain abhors boredom. The brain needs to be optimally stimulated; too much is bad, too little is just as intolerable. In the total absence of stimulus, the brain goes haywire — it hallucinates. This is why sensory deprivation is as much a torture as physical punishment. Most people consider exposure to a moderate degree of danger as "recreational," such as riding a roller coaster or engaging in rock climbing. Adventures like these heighten a person's consciousness and emotion by putting his nerves "on edge," whereas the subsequent release of tension and return to safety bring forth enormous gratification and relaxation. Art experience is a virtual adventure of the mind, paved with climaxes and anti-climaxes all the way.[50] Along with this mini-adventure is art's emotional cathartic function, as it helps to flush out the unwanted buildups in our unconscious, starting it anew on a tranquil base.

On the surface, art takes on many forms and genres. Visual art depends on light perception. Music depends on hearing. Each also incorporates other sensory and symbolic elements, such as the combination of melody and language in songs. When it comes to literature and drama, art becomes even more complex and multi-dimensional. Yet they all converge on the same end point — the evocation of feeling. As a product of emotion that is projected extra-corporeally, art detaches from its creator and lives its own existence, continuing to touch people in years and millennia to come. Thus, from so humble a gesture as a frog's calling, evolution has perfected a channel of emotional exchange that has become an essential part of humanity.

Lastly, let me add that art can be used as a vehicle to carry other messages. In commercial art, it entices a person to buy; in political art, it propels the audience to act; and in religious art, it brings comfort and instills a sense of fear and awe. But all these utilities are built on the ability of art to stir up human feeling. Poor art, for instance, makes for poor propaganda.

Notes and References

1. Bryson B. (2003) *A Short History of Nearly Everything*, Broadway Books, Chapter 26, p. 397. (Quoted with permission from Penguin Random House)
2. The word "feeling" is commonly used in four different ways: (1) tactile sensation as in "I feel my way in the dark"; (2) bodily sensation such as "I feel hot"; (3) intuitive or subjective knowledge as in "I feel the situation is bad"; (4) inner experience of an emotion. In this book only the last meaning is adopted.
3. The other brain structure that comes singly is the pituitary gland, which lies at the bottom of the brain; it seems the lowly location of the pituitary was less appealing to Descartes as worthy of the soul.
4. James W. (1890) *The Principles of Psychology*, vol. 2, Henry Holt; republished by Dover, New York, 1950, p. 451.
5. James W. *Ibid.*, vol. 2, p. 450.
6. Arnold MB. (1960) *Emotion and Personality*, Columbia Univ. Press, New York; Damasio AR. (1995) Toward a neurobiology of emotion and feeling: Operational concepts and hypotheses. *The Neuroscientist* **1**: 19–25.
7. Damasio AR. (1994) *Decartes' Error*, Putnam's Sons, New York.
8. Bard and Cannon demonstrated that the hypothalamus alone, when disconnected from the rest of the brain, is able to express emotional outpouring when stimulated. See: Cannon WB. (1927) The James-Lange theory of emotion: A critical examination and an alternative theory. *Am J Psychology* **39**: 106–124.
9. Kluver H, Bucy PC. (1939) Preliminary analysis of the functions of the temporal lobes in monkeys. *Archives Neurol and Psychiatry* **42**: 979–1000.
10. Olds J, Milner P. (1954) Positive reinforcement produced by electrical stimulation of septal area and other regions of rat brain. *J Comp Physiol Psychology* **47**: 419–427.
11. The mammalian brain comprises an outer thin layer of gray matter (cortex) made up of neuronal cell bodies and an inner mass of white matter consisting of nerve fiber tracts wrapped in myelin sheaths. The word "nucleus" (plural "nuclei") in neuro-anatomy refers to a pocket of gray matter (neuronal cell bodies) embedded deep in the white matter of the brain.
12. Volz H-P, Rehbein G, Triepel J, *et al.* (1990) Afferent connections of the nucleus centralis amygdalae. *Anat and Embryol* **181**: 177–194.

13. Sigurdsson T, Doyère V, Cain CK, LeDoux JE. (2007) Long-term potentiation in the amygdala: A cellular mechanism of fear learning and memory. *Neuropharmacology* **52:** 215–227.
14. Costa VD, Lang PJ, Sabatinelli D, *et al.* (2010) Emotional imagery: Assessing pleasure and arousal in the brain's reward circuitry. *Human Brain Mapping* **31:** 1446–1457; Sabatinelli D, Bradley MM, Lang PJ, *et al.* (2007) Pleasure rather than salience activates human nucleus accumbens and medial prefrontal cortex. *J Neurophysiol* **98:** 1374–1379.
15. Kaczmarek HJ, Klefer SW. (2000) Microinjections of dopaminergic agents in the nucleus accumbens affect ethanol consumption but not palatability. *Pharmacol Biochem Behavior* **66:** 307–312.
16. Berns GS, Moore SE. (2011) A neuronal predictor of cultural popularity. *J Consumer Psychology*; DOI: 10.1016/j.jcps.2011.05.001; quoted by *Science*, **332:** 1367 (2011).
17. Menon V, Levitin DJ. (2005) The rewards of music listening: Response and physiological connectivity of the mesolimbic system. *NeuroImage* **28:** 175–184; Salimpoor VN, van den Bosch I, Kovacevic N, *et al.* (2013) Interactions between the nucleus accumbens and auditory cortices predict music reward value. *Science* **340:** 216–219.
18. Bartels A, Zeki S. (2000) The neural basis of romantic love. *Neuroreport* **11:** 3829–3834; Acevedo BP, Aron A, Fisher HE, Brown LL. (2012) Neural correlates of long-term intense romantic love. *Scan* **7:** 145–159.
19. Fisher HE, Brown LL, Aron A, *et al.* (2010) Reward, addiction, and emotion regulation systems associated with rejection in love. *J Neurophysiol* **104:** 51–60.
20. Smith KS, Tindell AJ, Aldridge JW, Berridge KC. (2009) Ventral pallidum roles in reward and motivation. *Behav Brain Res* **196:** 155–167; Kringelbach ML, Berridge KC. (August 2012) The joyful mind. *Scientific American*, **307:** 40–45.
21. Butcher KS, Cechetto DF. (1998) Receptors in lateral hypothalamus area involved in insular cortex sympathetic responses. *Am J Physiol* **275** (*Heart Circ Physiol* **44**)*:* H689-H696.
22. Adolphs R. (2010) Emotion. *Current Biol* **20:** R549-R552.
23. Naqvi NH, Rudrauf D, Damasio H, Bechara A. (2007) Damage to the insula disrupts addiction to cigarette smoking. *Science* **315:** 531–534.

24. Vogt BA, Rosene DL, Pandya DN. (1979) Thalamic and cortical afferents differentiate anterior from posterior cingulate cortex in the monkey. *Science* **204:** 205–207.
25. Bechara A, Tranel D, Damasio H. (2000) Characterization of the decision-making deficit of patients with ventromedial prefrontal cortex lesions. *Brain* **123:** 2189–2202; Fellows LK, Farah MJ. (2007) The role of ventromedial prefrontal cortex in decision making: Judgment under uncertainty or judgment per se? *Cerebral Cortex* **17:** 2669–2674.
26. Paulus MP, Frank LR. (2003) Ventromedial prefrontal cortex activation is critical for preference judgments. *Neuroreport* **14:** 1311–1315.
27. Zeki S, Romaya JP, Benincasa DMT, Atiyah MF. (2014) The experience of mathematical beauty and its neural correlates. *Frontiers in Human Neurosci* **8:** 68.
28. Bandler R, Shipley MT. (1994) Columnar organization in the midbrain periaqueductal gray: Modules for emotional expression? *Trends in Neurosci* **17:** 379–389.
29. To be exact, Sherrington's interoception includes proprioception, the sensation of our body posture and extremities, as well as that of the internal organs. See: Sherrington CS. (1907) On the proprioceptive system, especially in its reflex aspect. *Brain* **29:** 467–482.
30. Craig AD. (2002) How do you feel? Interoception: The sense of the physiological condition of the body. *Nat Rev Neurosci* **3:** 655–666.
31. Cechetto DF, Calaresu FR. (1985) Central pathways relaying cardiovascular afferent information to amygdala. *Am J Physiol* **248:** R38-R45; Volz H-P, Rehbein G, Triepel J, *et al*. Afferent connections of the nucleus centralis amygdalae: A horseradish peroxidase study and literature survey. *Anat Embryol* **181:** 177–194.
32. Fischetti M. (Feb 2011) Your brain in love. *Scientific Am* **304:** 92.
33. Barik J, Marti F, Morei C, *et al*. (2013) Chronic stress triggers social aversion via glucocorticoid receptor in dopaminoceptive neurons. *Science* **339:** 332–335.
34. Alia-Klein N, Goldstein RZ, Kriplani A, *et al*. (2008) Brain monoamine oxidase-A activity predicts trait aggression. *J Neurosci* **28:** 5099–5104.
35. Miller A, Kennedy MA. (2011) MAO-A abuse exposure and antisocial behavior: 30-year longitudinal study. *British J of Psychiatry* **198:** 457–463.

36. Schackter S. (1964) The interaction of cognitive and physiological determinants of emotional state. In: Berkowitz L. ed. *Advances in Experimental Social Psychology* **1:** 49–80. Academic Press, New York.
37. Whalen PJ, Rauch SL, Etcoff NL, et al. (1998) Masked presentations of emotional facial expressions modulate amygdala activity without explicit knowledge. *J Neurosci* **18:** 411–418; Whalen PJ, Kagan J, Cook RG, et al. (2004) Human amygdala responsivity to masked fearful eye whites. *Science* **306:** 2061.
38. Hume D. (1740) *Treatise of Human Nature*, Book 3.
39. Wundt W. (1907) *Outlines of Psychology*, Wilhelm Englemann, Leipzig.
40. Zayonc RB. (1980) Feeling and thinking: Preferences need no inferences. *Am Psychologist* **35:** 151–175.
41. Danziger S, Levav J, Avnaim-Pesso L. (2011) Extraneous factors in judicial decisions. *Proc Natl Acad Sci USA* **108:** 6889–6892.
42. Villemure C, Bushnell MC. (2002) Cognitive modulation of pain: How do attention and emotion influence pain processing? *Pain* **95:** 195–199.
43. Personal communication.
44. Loriaux AL, Roitman JD, Roitman MF. (2011) Nucleus accumbens shell, but not core, tracks motivational value of salt. *J Neurophysiol* **106:** 1537–1544.
45. Among the modern philosophers, Suzanne Langer (1895–1985) devoted a great deal of effort to the topic of art. She correctly identified art as a vehicle of emotion and as a product of the brain. Nevertheless, her emphasis of art as symbol (like language) sidetracked the real physiological underpinnings of art. For details, see: Langer SK. (1942) *Philosophy in a New Key.* Harvard Univ. Press, Cambridge, MA; (1953) *Feeling and Form.* Scribner's Sons, New York; (1967/72) *Mind: An Essay on Human Feeling.* Johns Hopkins Univ. Press, Baltimore, MD. Langer subsequently attempted to correct this error; see: Langer SK. (1957) *Problems of Art.* Scribner's Sons, New York, p. 124.
46. Sigmund Freud provided a psychoanalytic explanation of art as a sublimation of suppressed erotic desire. Although Freud's theory does not contradict the emotional origin of art, it is far too narrow in scope and is applicable only to the human species. See: Freud S (1905) *Three Essays on the Theory of Sexuality.* Reprinted in 2000 by Basic Books, New York.

47. Menon V, Levitin DJ. (2005) The rewards of music listening: Response and physiological connectivity of the mesolimbic system, *Neuroimage* **28**: 175–184; Berns GS, Moore SE. (2012) A neural predictor of cultural popularity. *J of Consumer Psychology* **22**: 154–160, quoted by *Science* **332**: 1367 (2011); Salimpoor VN, van den Bosch I, Kovacevic N, *et al.* (2013) Interactions between the nucleus accumbens and auditory cortices predict music reward value. *Science* **340**: 216–219.
48. Kawabata H, Zeki S. (2004) Neural correlates of beauty. *J Neurophysiol* **91**: 1699–1705; Lacey S, Hagtvedt H, Patrick VM, *et al.* (2011) Art for reward's sake: Visual art recruits the ventral striatum. *Neuroimage* **55**: 420–433.
49. Ishizu T, Zeki S. (2011) Toward a brain-based theory of beauty. *PLOS ONE* **6**: e21852.
50. Obviously I am talking about arts with a time factor (music, literature, etc.), but even among the graphic arts, careful analysis with eye tracking can reveal sequential elements in their appreciation.
51. Cannon WB. (1932) *The Wisdom of the Body*. Norton, New York.
52. Bard P, in C. Murchison (ed.): Foundations Exp. Psycho., Clark Univ. Press, 1929, p. 251. There is new evidence that the sacral autonomic outflow is sympathetic in nature. Nevertheless, the new finding does not alter the theme of this chapter. See: Espinosa-Medina I, Saha O, Boismoreau F, *et al.*, (2016) The sacral autonomic outflow is sympathetic, *Science* **354**: 893–897.
53. Ferenczi EA, Zalocusky KA, Liston C, *et al.* (2016) Prefrontal cortical regulation of brain circuit dynamics and reward-related behavior. *Science* **351**: 41.

Chapter 10: *Self* and Memory

Memory is the paint applied on the canvas of your self-portrait.

Overview: *Memory is the stuff out of which the biographical self is built. Much of our personal identity depends on where we came from, what we have done, and what has happened to us. Without a reliable memory, much of the self is lost. Memory enhances fitness for survival as past experience influences future action in a way beneficial to the organism. The consolidation of memory involves synaptic modifications and requires activation of genes and protein synthesis. In higher animals the brain structure called the hippocampus plays a crucial role in converting short-term to long-term memory. Long-term changes in hippocampal neurons are detectable by electrical recording as sustained firing. Some neurons are involved in space memory, while others can be correlated with the sense of time.*

10.1 Memory and the Making of Biographical *Self*

In December of 2013, a man turned up in a snowdrift in Oslo, Norway. He spoke five languages and appeared intelligent, but he could not identify himself when questioned by Oslo Police. He carried no ID card and could not provide his name, age, address, and nation of origin, and had no idea what had happened to him. Four months later the police released his picture to the public, and only then was he recognized by his parents who lived in the Czech Republic. This true story was reported by NBC News.[1] How the man ended up one thousand miles away from home is irrelevant to this book, but the case illustrates the importance of memory in constructing the biographical *self*, the stuff of who a person is.

Such extreme conditions of memory loss are very rare, but more common and slow versions befall many elderly people suffering from Alzheimer's disease and vascular dementia, the latter caused by impaired blood flow to the brain or cumulative small strokes.[2] These people irreversibly lose their memory bit by bit, until they no longer are their own *self*. In the book bearing the depressing title *Remind Me Who I am, Again!* Linda Grant described how her demented mother had lost her recent memories and was about to lose the remote memories: "Only the deep past remained, which emerged at moments, in bits and pieces. ... This moment, the one she is really living in, is lost from sight as soon as it happens. And the long-ago memories are vanishing too. Only fragments remain. Soon, she will no longer recognize me, her own daughter. ... Memory, I have come to understand, is everything, it's life itself."[3]

Memory is the thread that connects the pieces of your past into a sensible whole. In a shorter version, it is the equivalent of a calling card you present to a new acquaintance at a cocktail party; in a longer version, it is the résumé that you prepare for a prospective employer. If your *self* is a portrait, memory is the paint applied on the canvas, one stroke at a time.

From the biological standpoint, animals need memory so that their experiences can influence future actions in a positive way — increasing their chance of survival. Unlike other animals whose duration of memory storage is limited by their life span, humans have the unique ability to preserve and transmit their memory across many generations in the form of writing and digitized records.

10.2 The Story of HM and the Secret to Human Memory

The first clue to unlocking the secret to human memory came in 1953, when a 27-year-old man named HM (Henry Molaison, 1926–2008) underwent surgery to cure his intractable seizures. The procedure consisted of removing his medial temporal lobe (mainly the hippocampus and its adjacent area entorhinal cortex) from both brain hemispheres.[4] The operation stopped the seizures and HM retained his normal intelligence and perceptual ability, and had a decent vocabulary. But, alas, he could not remember anything new since surgery! HM had developed what we now call *anterograde amnesia*, the inability to make new memories. For example, he could not recognize persons whom he routinely saw on a monthly basis, he could not retain information about an object for more than a few minutes, and he had great difficulty learning to get around a new house. This incident established the role of the medial temporal lobe in human memory.[5] The kind of memory lost in HM was *episodic* memory, knowledge of events with a time-place reference, such as "I had dinner last night with Mr. Smith in an Italian restaurant on Fifth Avenue." On the other hand, *semantic* memory, which concerns general factual statements such as "the Earth is round," was less affected. Episodic and semantic memories both belong to the so-called *declarative* (or *explicit*) memory, one that requires the full participation of consciousness. The opposite type of memory, known as *non-declarative* (or *implicit*) memory, which consists of training for motor skills and other types of learning that do not necessitate conscious effort to recall, was normal in HM. Motor skill learning depends on brain regions outside of the medial temporal lobe, and involves structures such as the basal ganglia and cerebellum. What is also significant about HM's memory is that events from the remote past were generally (but not totally) preserved, such as his childhood experiences. The lesson from HM is that the hippocampus is important for converting short-term to long-term memory of the declarative type. Figures 10.1 to 10.3 show the medial temporal lobe of the human brain and the parts relevant to memory, while Tables 10.1 and 10.2 compare the various types of memory.

Long-term memories are those that last for months and years after the experience; they are correlated with permanent physical changes in the brain, in particular the neurons and their synapses in multiple cortical areas. Short-term memories are those that stay in the brain for less than an hour before permanent physical changes set in. In HM, it lasted

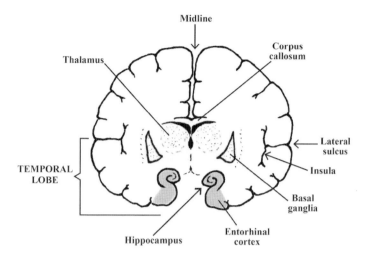

Fig. 10.1. Simplified drawing of a human brain in coronal section (vertical cross-section) showing the location of the hippocampus and entorhinal cortex (shaded areas) in both hemispheres. Note that the hippocampus is the in-rolling portion of the medial side of temporal lobe. Other brain areas are labeled for landmark purposes.

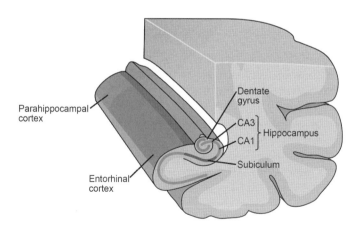

Fig. 10.2. Cutout view from the right half of Fig. 10.1 showing details of the hippocampus and related structures. The "hippocampus proper" or simply "hippocampus" comprises the CA3 and CA1 structures, whereas the "hippocampal formation" includes, in addition, the dentate gyrus and subiculum. Some authors consider the entorhinal cortex as part of the hippocampal formation, since it serves as the gateway to the hippocampus.[See Note 11: Kandel *et al.* courtesy McGraw-Hill.]

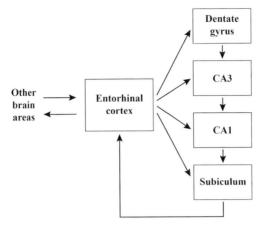

Fig. 10.3. Diagram of connections among the hippocampal components. See preceding figure for anatomical locations.

Table 10.1. Types of Memory According to Content

A. Declarative (explicit; requires conscious learning and recall; hippocampus-dependent)
  1. Episodic: With time/place reference (specific incidents)
  2. Semantic: Without time/place reference (general facts)
  3. Mixed: Many memories are a mixture of both. Biographical *self* depends on episodic memory for accuracy, but it also has semantic components.

B. Non-declarative (implicit; procedural; skill acquisition; learning may or may not need conscious effort; retrieval may be subconscious or minimally conscious; hippocampus-independent)
  1. Associative: Depends on correlation between two stimuli: conditioned stimulus (CS) and unconditioned stimulus (US). Prime example is simple classical conditioning of Pavlov. Operant conditioning (Skinner Box) is a variant in which conscious participation is needed.
  2. Non-associative: Priming; motor training or skill acquisition (e.g. riding a bicycle); habituation; sensitization or desensitization.

Note: "Priming" is a type of learning in which a prior single exposure to a stimulus facilitates subsequent learning with respect to the same stimulus. It works at the perceptual level. (See Note 46.) A standard way to test "priming" is to provide a subject with a certain nonsense word. Subsequent presentation of a panel of nonsense words will show that the subject is more likely to pick up the one he has been exposed to.

Table 10.2. Types of Memory According to Duration

A. Short-term: Usually less than an hour. Message resides in ongoing interactions between the prefrontal cortex (with involvement of other neocortical areas), and the hippocampus. Involves transient synaptic functional changes without permanent structural alteration.

B. Long-term: At least a day but usually months and years, up to a lifetime. For declarative memory, hippocampus serves as a way station for transfer of memory traces to distributed neocortical areas for permanent storage, where long-lasting structural changes take place. The transition process involves gene activation and protein synthesis. Recall depends on conscious executive effort of the prefrontal cortex in mobilizing memory traces stored in multimodal brain areas, including the associative cortical centers such as the parietal cortex.

Note: Both declarative and non-declarative memories can be short-term or long-term.

---

only two minutes. Short-term memory can be compared to a notepad while long-term memory is like a filing cabinet. The conversion from short- to long-term memory is called memory consolidation or fixation, and is hippocampus-dependent.[6]

Amnesia is defined as the loss of memory, either from a deficit in storing information or in recall. Amnesia due to a storage problem is called *anterograde* amnesia, whereas that due to defect in retrieval is called *retrograde*. HM's amnesia was primarily anterograde as he could not form any new memory after surgery. In retrograde amnesia, a person cannot remember what happened in the past. Many people recovering from head trauma cannot recall the event immediately before the impact, as in an automobile accident. More severe cases of retrograde amnesia extend far into years past. In HM's case, many (but not all) of his childhood memories were preserved after brain surgery.

Amnesia can also be classified as global or focal. A global amnesia covers all types of experience. However, even with the global type, some categories of experience are more severely affected than others. For example, time is usually lost first, followed by place, then by person. An Alzheimer's patient will first forget the time ("what day of the week is today?"), then the place ("where are you now?" or "Where do you live?"), and lastly, persons ("Have you seen this man before?"). This is understandable, since time moves forward and changes constantly. Places remain the same but the subject moves around and can get confused with space orientation. People's faces are remembered much longer as they change very slowly. Thus, inability to recognize people is a sign of advanced dementia. In very severe cases, even *self* is no longer recognized.

In the disorder called "transient global amnesia", an otherwise normal person (usually elderly) can suddenly lose all memories and not recognize family members. Luckily this condition lasts only a few hours, after which the past gradually returns, but what happens during the episode stays blank forever.

In focal amnesia, a block of memory within a given time span is obliterated (say, no recall of a period between five and ten years old), or the memory of a particular event is erased (when a soldier cannot recall a combat scene). Focal amnesia are often functional in nature, i.e., not accountable by a detectable structural damage to the brain. It is usually due to unconscious suppression, as in neurosis, hysteria, and post-traumatic stress disorder (PTSD). Hypnosis, of course, can induce reversible focal amnesia.

William James attributed the phenomenon of double personality to alternating lapses of memory. In his book, *Principles of Psychology*, he recounted a case of a certain Felida X, reported by Dr. Azam of Bordeaux: "At the age of fourteen this woman began to pass into a 'secondary' state characterized by a change in her general disposition and character, as if certain 'inhibitions', previously existing, were suddenly removed. During the secondary state she remembered the first state, but on emerging from it into the first state she remembered nothing of the second. At the age of forty-four the duration of the secondary state (which was on the whole superior in quality to the original state) had gained upon the latter so much as to occupy most of her time. During it she remembers the events belonging to the original state, but her complete oblivion of the secondary state when the original state recurs is often very distressing to her, as, for example, when the transition takes place in a carriage on her way to a funeral, and she hasn't the least idea which one of her friends may be dead. She actually became pregnant during one of her early secondary states, and during her first state had no knowledge of how it had come to pass. Her distress at these blanks of memory is sometimes intense and once drove her to attempt suicide."[7] It looks as if this lady had two biographical selves, based on two sets of memories that were partially overlapped.

10.3 Memory, Imagination, and Fantasy

Metaphorically, our conscious mind may be compared to the stage of a playhouse, on which only one play goes on at a time. In the default condition, our mind's "play" reflects what is going on in the immediate present. But when we recall, we are resetting the stage to play out a past event. To this end our mind mobilizes a limited repertoire of actors, costumes, props, and scenes (all deposited there by past experience) and uses them repeatedly in a modular manner. For example, the same chair can be employed in *The Death of a Salesman* and *A Streetcar Named Desire*; the same Laurence Olivier can play Hamlet and King Lear in a Shakespeare festival. However, the reconstruction requires simplification, abstraction, and reorganization, resulting in skewing and distortion, as commonly occurs when two witnesses testify the same crime scene. Many of us have the experience of returning to a childhood town, where we are stunned by the realization that the first school we attended was so small compared to the one we have harbored in our minds.

For those who prefer a "high-tech" metaphor for memory, our conscious mind can be likened to a computer monitor. Each person's mind has only one screen, but countless images can appear on the same screen by changing the values of the individual pixels and combining them in various patterns.

Imagination is similar to recall in that it too is a modular reconstruction of past experience, but it differs from memory as the parts can be fragmented and freely recombined without regard to real time and place. Fantasy enjoys even more freedom and may have contradictory or illogical contents. Nevertheless, some traces of reality remain. For instance, when I fantasize a Pegasus, both the horse and the wings are real but the combination is not logical. If fantasy appears involuntarily and is believed by the subject to be true, it becomes hallucination, a symptom commonly seen in schizophrenia.

That memory and imagination are closely related has been borne out in clinical neurology. Patients having lesions to the hippocampus not only lose their memories but also their ability to imagine. They produce only fragmented images lacking coherence.[8]

10.4 Animal Learning

Implicit (non-declarative) animal learning can be of non-associative or associative type. One example of non-associative learning is habituation, in which an animal learns to ignore a repeated stimulus if it turns out to be benign. A simple associative learning is the classical conditioning discovered by Pavlov early in the 20th century. During his research on the physiology of digestion in dogs, Pavlov observed that the dogs not only salivated in the presence of meat powder (a natural stimulus), but also in the presence of the laboratory technician who normally fed them, even in the absence of the food. Pavlov figured that the dog responded (salivation) to a neutral stimulus (person) by associating it with a natural stimulus (food). He then used a bell to call the dogs to their food and, after repetitions, the dogs started to salivate to the sound of the bell. In classical conditioning, a neutral stimulus (called *conditioned stimulus* or CS) is paired with and precedes a significant stimulus (called *unconditioned stimulus* or US), which normally elicits an innate, often reflexive, response (called *unconditioned response* or UR). After repetition, the two stimuli become associated and the animal responds (called *conditioned response* or CR) to the conditioned stimulus alone.

Operant conditioning is a paradigm of associative learning in which one particular stimulus interacts with a specific behavioral response, leading to the increase or decrease in the frequency of that behavior, depending on whether the stimulus is desirable (positive reinforcement) or undesirable (negative reinforcement) to the animal. Unlike in classical conditioning, in which a reflexive bodily response is modified by an antecedent stimulus, in operant conditioning a voluntary behavior is modified by a stimulus subsequent to the behavior. For example, the frequency of bar pressing (by a rat) increases when the action is rewarded by food. Conversely, the frequency decreases if the action triggers an electric shock. In a different setup, bar pressing can be designed to suppress an anticipated electric shock. Since operant conditioning involves manipulation of a gadget, it is also called *instrumental conditioning*. A well-known instrument for testing operant conditioning is the Skinner box. Unlike classical conditioning, which is strictly implicit in nature, operant conditioning has an explicit component.

Both classical and operant conditionings imply a rudimentary sense of causality in the experimental animal, though not necessarily at the conscious level.

Declarative (explicit) memory of the episodic type can be tested in lower animals, though it requires more elaborate and ingenious design. For example, rodents can be tested for space memory with the aid of visual cues. These include the Morris water maze and Barnes maze tests. The former requires an animal to swim in shallow water to reach a partially submerged platform, whereas the latter requires it to find a correct escape hole on a circular surface. Both depend on the integrity of the hippocampus.[9]

10.5 On Abstract Memory

No memory is strictly abstract. Memory is usually rooted in something concrete — experience related to sensation or musculoskeletal movement. Take telephone numbers for example. When I try to recall a phone number, I subconsciously use several cues at the same time, among them: (1) visual appearance of the digits on a piece of paper or in my mental "blackboard"; (2) tongue movements when calling out the number (remember the expression "on the tip of my tongue"?); (3) the sequential finger motion when dialing the number. I can recall on numerous instances the difficulty of saying out loud the number in Chinese if I initially stored it in English, although I am equally fluent in both languages. I also have difficulty using a rotary dial if the number is etched in my brain in the format of a rectangular keypad. These are examples of declarative memory aided by procedural memory. Another interesting fact concerns map memory. When going from one part of town to another, I rely on the east-west/north-south coordination on my mental map. My wife, by contrast, uses bodily movements as cues, like turning right followed by turning left, and so on — another use of procedural memory. My point is that abstract memories are usually tied to concrete references.

10.6 Emotion Enhances Memory

Emotion heightens alertness and facilitates memory fixation. Strong memories are those engraved with intense feelings. Remember where you were and what you were doing when President Kennedy was assassinated? Or, for the younger generations, when the World Trade Center was attacked on September 11, 2001? In our personal life, the indelible experiences are also those highly emotionally charged — the first erotic kiss, the first sexual intercourse, the first encounter with a dead person, the first scene of a newborn baby...

In animal training, there is evidence that during negative reinforcement nerve impulses from the hippocampus activate the amygdala (a fear center), whereas in positive reinforcement the hippocampus activates the nucleus accumbens (a motivation/reward center) instead.[10]

10.7 Cellular and Molecular Basis of Memory: Lessons From Invertebrates

The underlying mechanisms for all types of memories are quite similar. In a nutshell, short-term memories involve transient, functional changes in the synaptic connections, while long-term memories require, in addition, gene activation and protein synthesis, along with permanent structural alterations in the synapses.

From the early work of the Spanish anatomist Santiago Ramon y Cajal and subsequently the research of the British physiologist Charles Sherrington, the importance of synaptic connections as "switches" for the passage of information had been recognized. In 1949, the Canadian psychologist Donald Hebb proposed the "synaptic theory" of learning, stating that repeated firing across a synapse facilitates the future firing across the same synapse. However, demonstrating this process is a formidable task. Since each neuron is equipped with thousands of synapses, and there are 100 billion (1×10^{11}) neurons in the human brain, the total number of synapses per brain is astronomical, exceeding the number of stars in our galaxy. The brain, in a sense, is an enormously complex information-processing machine, with multiple parallel computations and re-entrant processing, along with endless feedback and feedforward loops. The only way to catch a glimpse of this picture is to seek a simple system, where interaction between single neurons can be demonstrated. Such a system was found in the invertebrate *Aplysia californica* (sea slug), a giant marine snail capable of exhibiting simple reflexes such as gill retraction as a defensive response when the body is touched (Fig. 10.4). Largely through the work of Eric Kandel and his associates, much cellular and molecular underpinnings of memory have been obtained by studying this model.[11]

The abdominal ganglion of *Aplysia* contains about 2,000 cells, a small number compared to the mammalian nervous system, many of which are large and identifiable. Those involved in gill withdrawal reflex, a defense response, are amenable to behavioral manipulations, including habituation, sensitization, and classical conditioning. Normally, lightly touching the siphon leads to gill withdrawal. Repeated touching (a harmless stimulus) leads to *habituation* (ignoring the stimulus). Electrical shock (noxious stimulus) applied to the tail followed by touching of the siphon facilitates the response to touch (called *sensitization*).

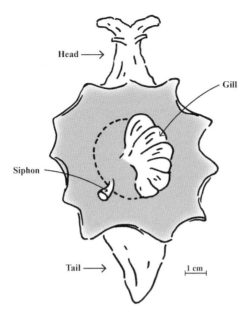

Fig. 10.4. Drawing of *Aplysia* showing the gill, the siphon, the head, and the tail.

Pairing of touching and electric shock (touch preceding shock) results in *classical conditioning* (Pavlovian conditioning), wherein touch alone without shock causes a strong gill withdrawal response. These behaviors can be demonstrated not only in the intact animal, but also in isolated neurons in tissue culture. Unlike habituation and sensitization, classical conditioning is an associative learning, where two stimuli are closely correlated to achieve the behavioral outcome. The three types of learning in Aplysia are summarized in Figure 10.5.

Kandel and his group found that habituation is correlated with a decrease, whereas sensitization with an increase, in the release of neurotransmitters across the synapse. In the gill withdrawal reflex (Fig. 10.5), the transmitter released by the sensory (touch) neuron "A" acting on the motor neuron "B" is glutamate, whereas the neuro-modulatory neuron "C" that conveys tail shock message releases serotonin on the pre-synaptic portion of the sensory neuron "A". Note that the connection between neuron C and A is "axon-axonic," serving to regulate the activity of the

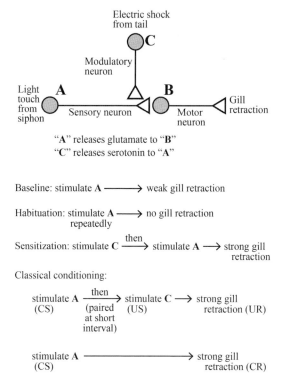

Fig. 10.5. Diagram showing the circuitry of three neurons involved in *Aplysia* gill retraction. Differences between habituation, sensitization and classical conditioning are shown. Under the baseline condition, when the siphon is lightly touched, the gill retracts weakly. On repeated touching, the response becomes weaker and weaker, until it ceases (habituation). If the tail is shocked, causing a strong retraction of the gill, subsequent touching of the siphon alone produces a strong retraction (sensitization). To establish classical conditioning, the animal must be trained by being subjected to paired stimulation of light touch (conditioned stimulus or CS) and electric shock (unconditioned stimulus or US), with touch preceding shock by a short internal. The unconditioned response (UR) of gill retraction reacting to shock becomes a conditioned response (CR) when, at the end of training, the conditioned stimulus (light touch) alone elicits strong gill retraction.[See Note 11: Squire & Kandel.]

"axon-dendritic" synapse between A on B. The modulatory function of C on A leads to the increased release of glutamate from A. The difference between sensitization and classical conditioning lies in the relation between the two stimuli (touch and shock). The former is non-associative and non-specific, as in a situation where any harmless noise will startle a person who just experienced bombing. In the latter, the first stimulus (light touch) must precede the second (shock) by a very short interval; once established, light touch alone (not any other mild stimulus) is capable of eliciting a strong gill withdrawal (now a conditioned response).[11]

As shown in Figure 10.6, classical conditioning incurs changes not only in the nerve terminal but also in the cell nucleus. First, synaptic activity triggers a series of events called signal transduction, which transmits the message to the nucleus to activate the gene, resulting in the syntheses of new messenger RNA and proteins, the latter being transported back to the same nerve terminal to strengthen the existing synapse and to make more synapses as well.[11] At the same time, changes also occur in the post-synaptic side in the receiving neuron, involving the glutamate receptors

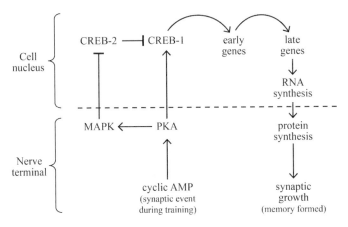

Fig. 10.6. Diagram showing gene activation in the nerve terminal (pre-synaptic component of a synapse) in Aplysia in the formation of long-term memory. During training, cyclic AMP in the cytoplasm activates PKA (protein kinase A, a signal transducer). The latter is then translocated to the nucleus to activate the transcription factor CREB-1, which induces the immediate *early* genes. The products of these genes in turn induce the *late* genes to produce proteins responsible for the permanent structural changes of the synapses, including growth of new synaptic connections. In the meantime, PKA also activates MAPK, another signal transducer, which inhibits the action of CREB-2, a suppressor of CREB-1. The net effect is the activation of CREB-1. These changes are universal across species, both invertebrates and vertebrates. The abbreviations are: CREB, cyclic AMP response element binding protein; MAPK, mitogen-activated protein kinase (also known as MAP kinase). [See Note 11: Squire & Kandel.]

(see Sec. 10.8 below with respect to vertebrates), resulting in the activation of the post-synaptic receptors and the growth of the dendritic spines. The net outcome is facilitation of transmission across the synapse.[12]

A newly proposed mechanism of *Aplysia* learning involves *prion* proteins. Prions are a type of proteins that exists in two conformations, the active and the inactive, the former being able to activate the latter in a self-sustained manner. A neuronal protein of the CPEB family that facilitates protein synthesis has been found to have prion-like properties. The conversion of CPEB in the synapses to the active form following learning is believed to contribute to memory storage by maintaining long-term synaptic changes. Once CPEB in a given synapse is activated, new proteins continue to be made and stored in a synapse-specific manner, providing a mechanism for "tagging" of memory trace to those synapses that have experienced learning.[13] Prion-like proteins have also been associated with long-term human memory.[14]

Another invertebrate model from which useful data on memory can be obtained is the fruit fly *Drosophila melanogaster*. In this animal, associative learning between odor and shock persists for 24 hours.[15] The process of memory consolidation can be tracked down to protein synthesis that occurs in two specific neurons.[16]

The roundworm *Caenorhabditis elegans* serves as another useful model for invertebrate memory, as it is capable of non-associative and associative learning. It undergoes habituation to mechanical and chemical stimuli, and can learn from smell, taste, temperature, and oxygen-level cues to the benefit of its survival. Specific genes underlying cellular and molecular mechanisms of learning have been identified.[17]

10.8 Cellular and Molecular Basis of Memory: Lessons From Vertebrates

In the vertebrates, much work has been carried out on the hippocampus of the rodent brain, both in live animals and in brain slices. One of the most remarkable observations that correlate with memory is

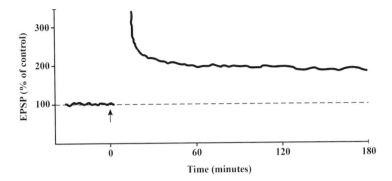

Fig. 10.7. Demonstration of long-term potentiation (LTP). Recording of excitatory post-synaptic potential (EPSP) from a hippocampal neuron of the rodent brain as a result of repeated stimulation of its pre-synaptic fiber. Four trains of 100 Hz stimulation (1 second duration, 5 minutes apart) given at zero time (arrow) leads to long-term sustained responsiveness that lasts for days or weeks, corresponding to long-term memory.

long-term potentiation (LTP), first reported by Lomo and Bliss in the 1970s, in which a burst of high frequency stimulation leads to long-lasting enhancement of excitatory post-synaptic potential (EPSP).[18] A recording of LTP is shown in Figure 10.7. The synaptic mechanism of LTP, explained in Figure 10.8, depends on the NMDA receptor and involves gene activation and protein synthesis in the post-synaptic neuron. This results in the growth (both size and number) of the dendritic spines and strengthening of synaptic connections.[19] LTP is believed to be universal for all animals.

LTP occurs in many parts of the brain other than the hippocampus. Further, the phenomenon of long-term *depression* (LTD), the opposite of LTP, has also been described. Synapses that exhibit LTP can also exhibit LTD, which is typically induced by sustained stimulation at low frequency (2 stimuli per second for 5 seconds).[20] Some neurons use LTD as a mechanism of learning, such as the cerebellar Purkinje cells.[21]

Another remarkable fact relevant to hippocampal function is the discovery of neurons that are capable of charting real-time location and physical orientation of the animal (rat), made in the 1970s by O'Keefe

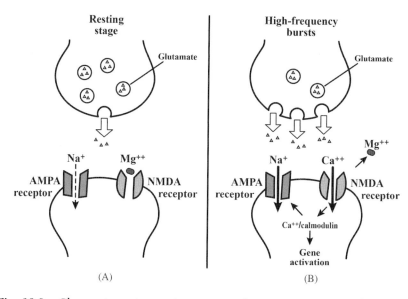

Fig. 10.8. Changes in post-synaptic neurons in long-term potentiation (LTP). Two types of glutamate receptors are depicted in the post-synaptic part of the synapse: the AMPA type receptor which allows sodium ion to enter, triggering action potential; the NMDA type receptor which regulates the function of AMPA receptor. (A) The **left panel** shows the resting stage when the NMDA receptor is blocked by magnesium (Mg^{++}). Under this condition, the neurotransmitter glutamate hitting the AMPA receptor causes only a small amount of sodium ion (Na^+) to enter the post-synaptic neuron. (B) The **right panel** shows that, upon high-frequency (tetanic) stimulation, the pre-synaptic neuron releases more glutamate. The strong depolarization of the post-synaptic membrane expels the magnesium and unblocks the NMDA receptor, allowing the influx of calcium (Ca^{++}). The calcium ion combines with the protein calmodulin (forming Calcium/calmodulin complex) and activates the enzyme calcium/calmodulin kinase II (CaMKII), which phosphorylates the AMPA receptor, facilitating the influx of sodium into the post-synaptic neuron, leading to transient potentiation. If the tetanic stimulation is repeated many times, calcium/calmodulin sends a message to the cell nucleus (through cyclic AMP, PKA, and MAPK), leading to gene activation and the synthesis of more glutamate receptors (both NMDA and AMPA) in the post synaptic neuron, along with the increase in the number of dendritic spines, resulting in synaptic facilitation. In addition, there is a feedback mechanism through which a retrograde message is sent from the post-synaptic to the pre-synaptic neuron, causing release of more glutamate. The abbreviations are: NMDA, N-methyl-D-aspartate; AMPA, alpha-amino-3-hydroxy-5-methyl-4-isoxazole-proprionic acid. For clarity and simplicity, intermediary steps from calcium/calmodulin to the AMPA receptor and cell nucleus are not depicted.

and his colleagues. This finding provides a mechanism for episodic memory, which, in essence, stores experiences in a *space–time* context. When live rodents are allowed to freely behave in a new environment, certain pyramidal cells in the hippocampus become active, as detected by intra-cerebral electrode recording. A cell will fire only when the animal is in a particular location and position, and many cells are recruited to form a cognitive map of the entire space. For each environmental setting, a specific cognitive map is formed that can later be used for recall.[22] What is interesting is that the storage of these space memories requires LTP in the neurons.[23]

Further studies have identified at least four types of pyramidal cells in the rodent hippocampal formation responsible for space memory: the *place cells* which fire when the animal is situated in a unique location; the *grid cells* which fire periodically in a grid-like pattern as the animal traverses at a given *speed*; the *head direction cells* which are active when the head is in a specific orientation regardless of the coordinate; and the *border cells* which react when a border is in the proximity.[24] Whereas *place cells* are located in the hippocampus proper, the other three types (grid, head direction, and border cells) fire most intensely in the entorhinal cortex, the gateway to the hippocampus.[25] Using the highly specific optogenetic study, there is evidence that signals from the three entorhinal cell types converge on the hippocampal pyramidal *place cells*, which integrate the input to generate a reliable *place field* for spatial memory. The *speed* sensitivity of *grid cells* further adds a time element to the *place field*. Thus, *place cells* have access to both self-motion and landmark-based information during navigation.[26] The importance of *place cells* in the formation of *abstract* memory has also been suggested.[27]

More direct evidence for the recording of *time* in the hippocampus has recently been obtained. When rats were trained in a two-part task with a 10-second gap in between, electrical recordings showed that many hippocampal neurons were sequentially active during this "silent" period, suggesting the encoding of time interval. Thus, the presence of "*time cells*" as well as "place cells" strengthens the role of hippocampus in episodic memory.[28]

It is important to ask whether the hippocampus is involved in recall other than encoding memory. Indeed, in a mouse study, it was demonstrated that stimulating a population of neurons in the dentate gyrus of the hippocampus is sufficient to trigger the recall of a fear experience, suggesting the presence of memory engram (coding) in the hippocampus.[29]

As a rule, adult neurons do not divide. One exception occurs in the dentate gyrus of the hippocampus (see Figs. 10.2 and 10.3), which continues to produce new neurons in the adult life. The ability of the hippocampus to undergo neurogenesis has been related to its ability to process memory. The dentate gyrus is the first relay station in the hippocampus for incoming information from the cerebral cortex. It is here where the information is split up and distributed among many cells in other parts of the hippocampus. The packaged information is then passed back to the cerebral cortex for long-term storage.[30] The newborn neurons in the dentate gyrus are more responsive than the older neurons to incoming excitatory input from the entorhinal cortex, providing evidence that neurogenesis is important for processing recent memory.[31] On the other hand, it has been suggested that, since the new neurons will have to integrate into the existing hippocampal circuits, too much neurogenesis may interfere with old memories already in place. Hence, a proper memory function depends on an optimum degree of adult neurogenesis.[32]

Most of the studies on the hippocampus were carried out in rodents. The question is whether some of the evidence is also found in human subjects. The answer is "yes." It was demonstrated, for example, that electrical stimulation of the human entorhinal cortex enhances the encoding of space memory.[33] Functional imaging of the human brain provided further evidence for the presence of *grid cells*.[34] During active retrieval of space-relevant memory, neurons in the human hippocampus that were originally encoded for the event are reactivated.[35] Thus, the storage of life events in a spatiotemporal context, a process essential for the building of autobiographical *self*, has a strong neurological basis.

10.9 Visible Morphological Changes Resulting From Learning

Following are some obvious morphological alterations correlated with memory: (1) Cab drivers in London who have to memorize the detailed city map end up having a larger hippocampus than the general population.[36] (2) Gross changes in brain areas are observed after learning music[37] or following the frequent use of a finger (Fig. 10.9).[38] (3) Increase in the number and size of dendritic spines are observed in specific areas of the brain following learning.[39] (4) In songbirds, neurogenesis goes on throughout life in the "vocal areas" of the brain, and the activity increases when the bird acquires a new song.[40]

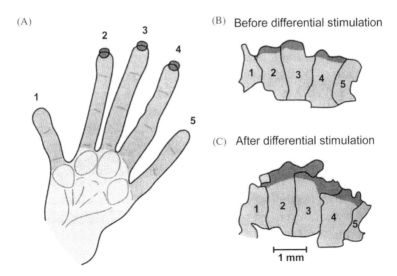

Fig. 10.9. Training increases cortical area representation. A monkey was trained to perform a task that required the use of the tips of fingers 2, 3 and 4. After a period, the cortical areas representing the stimulated fingertips became larger than others. A, location of the fingertips involved; B, cortical representation before training; C, cortical representation after training. [See Note 11: Kandel *et al.* courtesy McGraw-Hill.]

10.10 Dialogue between the Neocortex and Hippocampus

According to the systems theory of memory consolidation, declarative memory starts as transient changes in the conscious brain, assembling a variety of sensory inputs, using the prefrontal cortex as the coordination center and the parietal cortex as the associative center. Every day we receive countless information that involves time, place, emotion, intention, personal relations, and many others. For these fleeting experiences to become long-term memory, the information from widely distributed neocortical networks must be exported to the hippocampus, where they are filtered, extracted, categorized, and coded. In effect, the hippocampus fuses incoming information into a coherent trace. Consolidation consists of gradually returning the memory trace to the neocortex for permanent storage, by repeatedly activating the hippocampo-cortical pathway. This is accompanied by activation of the genes, first in the hippocampus, then in the cortical areas. This time-dependent storage process takes hours or days to accomplish, depending on the species and the emotional accompaniment of the experience. During this consolidation period the memory trace is vulnerable to disruption, either by drugs or interference with other experiences. As the memory trace "matures," the role of the hippocampus gradually (but not totally) recedes to the background, leaving the cortical areas alone to sustain the permanent memory. However, there is evidence that even long after memory consolidation, the hippocampus is still needed to keep the old memories intact.[41]

In real life we rarely learn things that are totally new. More often new information is mingled with pre-existing ones. To avoid perturbation of the old knowledge, deposition of new facts requires the retrieval of the old ones for integration and assimilation. This involves parallel encoding in both the neocortex and hippocampus. Therefore, every input from new experiences results in reorganization and strengthening of the old ones.[42] This scenario is consistent with the concept that hippocampal neurogenesis during learning results in the rewiring of the existing network.

10.11 Newton, Einstein, Kant and the Little Hippocampus

Space and time are properties of our universe that have baffled mankind since the beginning of history. In the classical mechanics of Newton, space and time are taken to be rigid and immutable frameworks within which matters are embedded, and space–time presumably remains unchanged even if devoid of matter and energy.[43] Contrary to this absolutist view, Einstein proposed that space–time is an inter-related four-dimensional continuum that changes with respect to the velocity of an observer. For example, a clock that travels at very high speed appears to tick slower to a stationary observer (time dilation), and the length of a fast-moving object appears shorter (in the direction of travel) to the same observer (length contraction). In Einstein's view, the perception of space and time is "in the eye of the beholder".[44] Kant took a bolder stance in dealing with this issue. Kant posited with his *a priori* notion that space and time are innate functions of the mind to make sense of the outside world. Although Kant's idea was never clearly formulated, the fact that he involved the participation of the observer had a hint of prescience.[45] But what does this have to do with the hippocampus, the seahorse-shaped structure of the brain that rolls inward out of sight from the surface?

From a micro-structural viewpoint, memory is a matter of synaptic plasticity, the strengthening or weakening of neuronal connections. From a macro-structural viewpoint, memory is a product of systems interaction among different parts of the brain where these microscopic changes take place. At any given moment the conscious brain engages multiple sensory modalities to perceive the external world, resulting in transient changes in various cortical areas and circuits. As soon as each fleeting moment of consciousness turns from the present to the past, the experience that is worth keeping is sorted and stored in a space–time coordinate. The hippocampus is the machine capable of this task. Once an immediate experience arrives, the hippocampus functions as a librarian and also as an archivist, who catalogues and stores the information, creating declarative memories that make up most of our biographical *self* — and the identity of who we are.

Notes and References

1. NBC News: How did amnesiac dubbed "John Smith" end up one thousand miles from home? By Henry Austin, April 10, 2014.
2. Dementia is defined as loss of cognitive function, the ability to make sense of and deal with the external world.
3. Grant L. (2000) *Remind Me Who I am, Again!* Granta Books, London. (Quoted with permission from Linda Grant.)
4. Hippocampus is the Greek word for sea horse, so named because of its shape. Unlike most other areas of the cerebral cortex which are made up of six-layered neurons (isocortex or neocortex), the hippocampus belongs to the evolutionarily older type of cortex and has only three layers of neurons (allocortex or archicortex).
5. Scoville WB, Milner B. (1957) Loss of recent memory after bilateral hippocampal lesions. *J Neurol Neurosurg Psychiatry* **20:** 11–21.
6. Some people use "working memory" to denote the "ultra-short" type of short-term memory, such as keeping a phone number in your head for the duration of dialing it. Nonetheless, the distinction between working memory and consciousness is blurred.
7. James W. (1890) *The Principles of Psychology.* vol. 1, Henry Holt; Dover, New York, 1950, p. 379.
8. Hassabis D, Kumaran D, Vann SD, Maguire EA. (2007) Patients with hippocampal amnesia cannot imagine new experiences. *Proc Natl Acad Sci USA* **104:** 1726–1731.
9. Morris RGM. (1981) Spatial localization does not require the presence of local cues. *Learning and Motivation* **2:** 239–260; Barnes CA. (1979) Memory deficits associated with senescence: A neurophysiological and behavioral study in the rat. *J Comp Physiol Psychol* **93:** 74–104.
10. Luo AH, Tahsili-Fahadan P, Wise RA, *et al.* (2011) Linking context with reward: A functional circuit from hippocampal CA3 to ventral tegmental area. *Science* **333:** 353–357.
11. Kandel ER, Schwartz JH, Jessell TM, *et al.* (2013) *Principles of Neural Science.* 5th ed. McGraw-Hill, New York; Squire LR, Kandel ER (1999) *Memory: from mind to molecules.* Scientific Am Lib New York.
12. Murphy GG, Glanzman DL. (1999) Cellular Analog of Differential Classical Conditioning in *Aplysia*: Disruption by the NMDA Receptor Antagonist DL-2-Amino-5-Phosphonovalerate. *J Neurosci* **19:** 10595–10602.

13. Si K, Lindquist S, Kandel ER.(2003) A neuronal isoform of the Aplysia CPEB has prion-like properties. *Cell* **115**: 879–891.
14. It has been reported that humans carrying a certain genotype of *prion* are associated with better long-term memory. See: Papassotiropoulos A, Wollmer MA, Aguzzi A, *et al.* (2005) The prion gene is associated with human long-term memory. *Human Molec Genetics* **14**: 2241–2246.
15. Quinn WG, Harris WA, Benzer S. (1974) Conditioned behavior in *Drosophila melanogaster. Proc Natl Acad Sci USA* **71**:708–712.
16. Chen C-C, Wu J-K, Lin H-W, *et al.* (2012) Visualizing long-term memory formation in two neurons of the Drosophila brain. *Science* **335**: 678–685.
17. Ardiel EL, Rankin CH. (2010) An elegant mind: Learning and memory in *Caenorhabditis elegans. Learning Memory.* **17**: 191–201.
18. Bliss T, Lømo T. (1973) Long-lasting potentiation of synaptic transmission in the dentate area of the anaesthetized rabbit following stimulation of the perforant path. *J Physiol* **232**: 331–356.
19. The events in the post-synaptic dendritic spine involve the NMDA type of glutamate receptor and the activation of signal transducers PKA and MAPK. These transducers enter the cell nucleus and activate the transcription factor CREB, which in turn leads to gene expression making possible synthesis of new proteins to be used for the dendritic spine. Note that the mechanism of post-synaptic gene activation is in many ways similar to the pre-synaptic counterpart depicted in Fig. 10.6. See: Impey S, Obrietan K, Wong ST, *et al.* (1998) Cross talk between ERK and PKA is required for Ca2+ stimulation of CREB-dependent transcription and ERK nuclear translocation. *Neuron* **21**: 869–883; Barco A, Alarcon JM, Kandel ER. (2002) Expression of constitutively active CREB protein facilitates the late phase of long-term potentiation by enhancing synaptic capture. *Cell* **108**: 689–703; Pittenger C, Huang YY, Paletzki RF, *et al.* (2002) Reversible inhibition of CREB/ATF transcription factors in region CA1 of the dorsal hippocampus disrupts hippocampus-dependent spatial memory. *Neuron* **34**: 447–462; Zhai S, Ark ED, Parra-Bueno P, Yasuda R. (2013) Long-distance integration of nuclear ERK signaling triggered by activation of a few dendritic spines. *Science* **342**: 1107–1111.
20. The mechanism of long-term *depression* (LTD) goes like this. Low frequency stimulation of the NMDA receptor leads to a partial relief of

magnesium blockade of the receptor, but the small calcium influx is insufficient to activate CaMKII. Instead, the small, sustained calcium flow activates protein phosphatase 1 (PP1), which, through removal of phosphates, leads to the internalization of AMPA receptor, weakening the synaptic strength. The outcome is the impairment of synaptic transmission, the opposite of LTP.

21. Koekkoek SKE, Hulscher HC, Dortland BR, *et al.* (2003) Cerebellar LTD and learning-dependent timing of conditioned eyelid responses. *Science* **301**: 1736–1739.
22. O'Keefe J, Dosrovsky J. (1971) The hippocampus as a spatial map: Preliminary evidence from unit activity in the freely-moving rat. *Brain Research* **34**: 171–175; O'Keefe J, Nadel L. (1978) *The Hippocampus as a Cognitive Map*. Oxford Univ. Press, New York.
23. Rotenberg A, Mayford M, Hawkins RD, *et al.* (1996) Mice expressing activated CaMKII lack low frequency LTP and do not form stable place cells in the CA1 region of the hippocampus. *Cell* **87**: 1351–1361.
24. The two main morphological types of hippocampal neurons are the pyramidal (triangular shaped) and the granule (round) cells.
25. Moser EI, Kropff E, Moser M-B. (2008) Place cells, grid cells, and the brain's spatial representation system. *Annual Rev Neurosci* **31**: 69–89.
26. Zhang S-J, Ye J, Miao C, *et al.* (2013) Optogenetic dissection of entorhinal-hippocampal functional connectivity. *Science* **340**: 44.
27. Memory of location (or "place") can be used as an aid to recall an abstract item, such as a number or letter. This strategy, called the "method of loci," was invented by ancient Greeks and Romans, in which symbols are embedded on different spots in a recognizable picture, like a landscape. Even today, students taking a test usually associate a number or statement to be memorized with its relative location in a concrete book — position in a volume and on a page. Thus the function of grid cells and place cells extends beyond the utility in navigation to include memory in general.
28. MacDonald CJ, Lepage KQ, Eden UT, Eichenbaum H. (2011) Hippocampal "time cells" bridge the gap in memory for discontiguous events. *Neuron* **71**: 737–749.
29. Liu X, Ramirez S, Pang PT, *et al.* (2012) Optogenetic stimulation of a hippocampal engram activates fear memory recall. *Nature* **484**: 381–385.

30. Clelland CD, Choi M, Romberg C, *et al*. (2009) A functional role for adult hippocampal neurogenesis in spatial pattern separation. *Science* **325:** 210–213.
31. Marin-Burgin A, Mongiat LA, Pardi MB, Schinder AF. (2012) Unique processing during a period of high excitation/inhibition balance in adult-born neurons. *Science* **335:** 1238–1242.
32. Frankland PW, Köhler S, Josselyn SA. (2013) Hippocampal neurogenesis and forgetting. *Trends in Neurosci* **36:** 497–503.
33. Suthana N, Haneef Z, Stern J, *et al*. (2012) Memory enhancement and deep-brain stimulation of the entorhinal area. *N Engl J Med* **366:** 502–510.
34. Doeller CF, Barry C, Burgess N. (2010) Evidence for grid cells in a human memory network. *Nature* **463:** 657–661.
35. Miller JF, Neufang M, Solway A, *et al*. (2013) Neural activity in human hippocampal formation reveals the spatial context of retrieved memories. *Science* **342:** 1111–1114.
36. Maguire EA, Gadian DG, Johnsrude IS, *et al*. (2000) Navigation-related structural change in the hippocampi of taxi drivers. *Proc Natl Acad Sci USA* **97:** 4398–4403.
37. Gaser C, Schlaug G. (2003) Brain structures differ between musicians and non-musicians. *J Neurosci* **23:** 9240–9245.
38. Jenkins WM, Merzenich MM, Ochs MT, *et al*. (1990) Functional reorganization of primary somatosensory cortex in adult owl monkeys after behaviorally controlled tactile stimulation. *J Neurophysiol* **63:** 82–104.
39. Engert F, Bonhoeffer T. (1999) Dendritic spine changes associated with hippocampal long-term synaptic plasticity. *Nature* **399:**66–70.
40. Goldman SA, Nottebohm F. (1983) Neuronal production, migration, and differentiation in a vocal control nucleus of the adult female canary brain. *Proc Natl Acad Sci USA* **80:** 2390–2394; Alvarez-Buylla A, Theelen M, Nottebohm F. (1988) Birth of projection neurons in the higher vocal center of the canary forebrain before, during and after song learning. *Proc Nat Acad Sci USA* **85**: 8722–8726; Nottebohm F. (Feb 1989) From bird song to neurogenesis. *Scientific Am* **260:** 74–79.
41. Steinvorth S, Levine B, Corkin S. (2005) Medial temporal lobe structures are needed to re-experience remote autobiographical memories: Evidence from H.M. and W.R. *Neuropsychologia* **43:** 479–496.

42. Tse D, Takeuchi T, Kakeyama M, *et al.* (2011) Schema-dependent gene activation and memory encoding in neocortex. *Science* **333**: 891- 895.
43. Newton I. (1686) *Philosophiae Naturalis Principia Mathematica.*
44. Lightman A. (2000) *Great Ideas in Physics.* McGraw-Hill, New York.
45. Kant I. (1781) *Critique of Pure Reason.*
46. Tulving E, Schacter DL. (1990) Priming and human memory systems. *Science* **247**: 301–306.

Chapter 11
Self and Free Will

Give me liberty, or give me death!

— Patrick Henry, 1775

Overview: (1) Objectively, free will is the ability of an animal to make choices or take alternative actions. Subjectively, freedom is an instinctive biological need that imparts a sense of agency for an action and is important for the assertion of self.

(2) Free will does make a difference to the world, to a limited extent. It does this by changing the probability of events, making them more likely, or less likely, to happen. Free will is biologically adaptive; it alters the environment for the benefit of the self.

(3) The brain is the convergent point of myriad causes. Every animal action is caused, but not every putative cause results in an immediate action. At any moment the brain serves as a selector by responding to some, but not all, causes. If the selector in the brain is of a physical nature, it is just part of the physical causal chain. At issue is how much the mind, if present and distinct from the brain, may influence the selection process.

(4) The physical world is an enormous, interlocking causal network, so much so that deterministic events can be difficult to predict and may come as a surprise. As a complex organ, the brain can generate behavior as difficult to predict as weather. If the mind does not exist, free will could be reduced to a series of physical changes in the brain that could be mistaken as volitional. However, if the mind does exist, one cannot rule out the possibility that free will interacts with the physical brain and contributes to the behavioral outcome. In the last analysis, the free will controversy remains open until the mind-body problem is settled.

(5) *If free will is "free," can it happen without a cause? Self, defined as a system that seeks its own perpetuation, is a basic principle permeating all life forms. The craving for perpetuation is a default cause of all human activities. Therefore, from the perspective of "self," free will as a human act always carries an implicit cause, no matter how "free" it appears.*

11.1 To Will or Not to Will?

Consider this scenario. Two men are walking on a road, debating whether free will is real. The first believes that people do have free will; the second denies it, insisting that it is just an illusion in a causal world. Then suddenly a boulder comes rolling down the road, giving them only two seconds to act. The first quickly escapes to safety by jumping to one side. What do you think the second will do? If he ignores the impending danger and does nothing, letting the event take its course, he will be killed. But if he also jumps out, he will be admitting that in this instance a little free will makes a difference between life and death.

We like to be in control of our own bodies, not just be puppets with invisible strings attached to our extremities. As William James said, "The whole sting and excitement of our voluntary life depends on our sense that in it things are really being decided from one moment to another, and that it is not the dull rattling off of a chain that was forged innumerable ages ago."[1] This morning I deliberately chose tea over coffee, and I believe this simple decision was not preconditioned by what my cave-dwelling forebears did twenty thousand years earlier, nor was it a remote consequence of quantum fluctuations in the beginning of time. To be sure, whereas the entire physical world appears to be locked in a deterministic chain, we humans seem to be able to disobey this ironclad predestination. Is this self-deception? In the objective sense, free will is the unpredictable element in a person's behavior as it appears to an observer. For example, a person can choose to act contrary to common sense, like giving away good food while going on starving. Looking at lower animals, a turkey can veer to the left or right unpredictably as you chase it for the Thanksgiving meal.

Humans enjoy many freedoms they take for granted until they are taken away. No one is happy to be told what to eat, what to wear, what to like and dislike, what to think, and how to act. Under the banner of freedom and liberty, countless people willingly face the firing squad or walk up to the gallows. What makes freedom so precious that some people value it above their own lives? Does freedom make biological sense?

I shall start from the simple premise that all animals take actions, directly and indirectly, *for the benefit of self* — their survival and reproductive success. Free will makes these actions possible, and therefore is favored by evolution. In this context I define free will as *the ability to initiate an action, in the absence of coercion, which entails a choice to act or not to act, and to elect which of the alternative courses of actions to take.* In the subsequent sections I shall examine free will from multiple perspectives. One topic I try to omit is the connection between free will and moral (or social) obligation and legal responsibilities, as I hold such discussions futile and irrelevant. Instead, in the next chapter I shall expound the evolutionary advantage of morality in the context of society as an expanded *self*.

11.2 When Is an Action Free?

My opinion is that since being free is a subjective feeling, any attempt at an objective explanation is likely to be inadequate.[2] One could say that an action is free if the agent could have done otherwise, in other words, if he has a choice. But making a choice is not always free. For example, if a person is held at gunpoint to surrender his money, he still can choose to resist and be killed. He has a choice, but no common sense would consider that a normal choice, since it is against biological instinct and it is unlikely that he intends that to happen. My point is that an action is free if it gives the executor a *sense of agency*. The sense of agency stems from the ability to *fulfill an urge to act, and the satisfaction of having it carried out*. (It should be noted that an urge not to act when conditions are conducive to act, and an urge to suppress another urge, both qualify as free will in this context.) Much suffering ensues when the impulse to act is constrained by external forces, as is evident in a howling pig when its four legs are tied, or in a defiant puppy put on leash for the first time. Humans at both the individual and collective levels strive to have this sense of fulfillment. We see this in an angry mob demonstrating for "freedom of expression" — whatever they wish to express is a different matter. The sense of agency promotes the affirmation of *self*. It is a basic biological need on par with feeding and sex. An animal needs freedom as much as an empty stomach needs food. This is why all animals, fish and philosophers included, enjoy freedom.[3,4]

11.3 Causation, Determinism, Unpredictability, and Surprise

The relation between determinism and free will is a tricky one, and has been a topic of debate for hundreds of years. If on the one hand we take the hard line by claiming that everything happening in the world at any moment is the consequence of what happened earlier, and if on the other hand we take the position that free will has the power to change the natural course of the world, there would be no compromise between the two (the incompatibilist view). One of them will have to give way, or perhaps both will have to be modified in order to mutually accommodate. Is this possible?

Our concept of determinism springs from the sense of causation in our daily life. As pointed out by David Hume, two events are causally related if one is always (empirically) followed by the other, such as A is followed by B. Conversely, if A does not occur, B will not happen, assuming all other causes of B are held constant. The relationship is even more convincing if it can be quantified. For example, if one unit of force can push a ball forward by five feet, then two units of force will push it by ten feet. In science, causality between two events is established by holding all other variables constant. From such observations science formulates physical laws, which govern many causal events of the inanimate world to a high degree of predictability. Newton's laws of motion are prime examples of determinism,[5] giving us a sense of clockwork precision, leading to the proclamation by Laplace (1749–1827) that, if he could know the positions and movements of all the heavenly bodies at one moment, he would be able to predict the outcome of the universe well into the future *ad infinitum*. Of course, Laplace's assertion is unverifiable because no one can ever have complete knowledge of the universe at any given moment. This boils down to the issue of complexity. Although isolated events are easy to predict, a complex network of causal chains frequently thwarts our forecast. And the world is an enormous, entangled causal network.

Let me start with a simple example. Suppose I stand a sharp pencil on end. Can I predict to which direction it will fall? In principle, a perfectly balanced pencil should never fall. But in real life it will. The standing pencil is under the influence of innumerable forces other than gravity: inclination and friction of the table surface, air flow and temperature fluctuations in the room, and building vibration due to wind speed or rumble of a passing train. Other distant factors may include an earthquake in a remote corner of the continent, plate tectonic movement below the ground, wobbling of the rotating Earth, the pull of the moon, or explosion of a distant star. Some factors are so weak that they are mere "noises," but they will show up when the major factors cancel out. So the pencil will fall, no matter how well I balance it. The more perfectly I try to control the variables, the more unpredictable will be the direction of the fall.

Take the case of weather forecasting. In the model called *chaos theory* (deterministic chaos), a nonlinear dynamic system can generate random, non-periodic, widely divergent outcomes that are hard to predict, depending on minute differences in the initial conditions. First hinted at by the mathematician Henri Poincare, it was discovered in the twentieth century by Edward Lorenz when he tried to solve three simple equations for the motion of the atmosphere.[6] The weather predicted by a computer varied widely when initial values differed by less than 0.1%. Thus, a butterfly flapping its wings in South America can potentially create a storm in Asia. This is known as the "butterfly effect." Although theoretically the event is entailed in the mathematical equations, in practice the outcomes are highly unpredictable since no one can be sure of the initial conditions at any given moment. The weather bureau gives up forecasting for more than ten days because that would require measurements to a degree of accuracy unattainable.

Another example of a deterministic condition that produces a surprising outcome is the *catastrophe theory*, a mathematical model in which a minor alteration in one variable at the critical point leads to an abrupt, high magnitude change in another. In simple words, a continuous action may produce an unexpected, sudden, discontinuous outcome.[7] The condition is said to be *metastable*. We see this when the addition of a single grain of sand brings forth the collapse of a sand pile, or when the passing of a casual skier precipitates an avalanche of an entire slope.

The bottom line is, although natural events are deterministic, in the real world they are not always predictable, and frequently may come as a surprise. Some philosophers take this as evidence that free will is just an illusion.[8] In the subsequent sections I shall explain why this extreme stance is not justified.

11.4 Navigating a Probabilistic World

We usually communicate about the future using such terms as "most likely," "more likely," "less likely" or "highly unlikely." Sometimes we are more definite and we may say that this or that "will surely happen" or "will surely not happen." These are all semi-quantitative terms to characterize a probabilistic world. The weather bureau tells us in a more scientific way that there is, say, a 50% or an 80% chance of rain. In mathematics, probability is expressed as the ratio of the number of expected occurrences to the total number of possible occurrences, and it ranges from 0 to 1.00, with 0 as no chance and 1.00 as definite to occur. Most future events fall within the two limits. Thus, an 80% chance of rain will mean a probability of 0.80. As the saying goes, "never say never." The future is open-ended, and events turn into certainty only at a juncture we called "the present." What accumulates in the past is a stockpile of certainties.

Natural laws are derived by observing the regularity of events using a process called induction. Inductive confidence increases with the number of observations made, or the size of the population observed; it approaches but never reaches certainty. For illustration, if you inspect 100 apples, and they are all good, you may conclude with some confidence that all the rest are good, but it may turn out that the next one (the 101st) is bad. Next, if you inspect 1,000 apples, and they are all good, you may conclude that all the rest are good, this time with even more confidence, but again it may turn out that the next (the 1001st) is bad. David Hume in the eighteenth century expressed doubt about the reliability of induction. In a more humorous tone, Bertrand Russell presented the parable of the inductivist chicken. The chicken, being fed by the farmer every day, predicted that it would continue to be fed. With each passing day it strengthened its inductivist belief, only to be surprised when one morning the farmer came and wrung its neck.[9] Russell concluded that certitude does not belong in this world.

Whether we like it or not, we live in a probabilistic world. We cannot sort out the multitude of interacting factors in nature. The overall result is seen in the variation of individual occurrences, be it a leaf, a flower, a snowflake, or a person. To deal with uncontrollable variations at different levels of reality, the mathematic science of statistics was invented. Statistics deals with an entire group by assuming each occurrence as a *random* event, regardless of the underlying mechanism. It gives us the *mean* (average) of the group, along with the degree of confidence in the form of *standard deviation*. Statistics is the only way to make sense of a probabilistic world.

We see this approach applied to a bag of gas. Molecules in a gaseous state are in constant heat motion, called Brownian motion. Although presumably each molecule jostles around according to prescribed mechanical laws, it is impossible to track down all of them individually. Boltzmann had the insight of treating them statistically (called statistical mechanics) as a collection of randomly moving particles, a concept called the *kinetic theory of gases*.[10] In this manner the unruly behaviors of the particles become manageable, and the energy level can be calculated.

11.5 The Brain as a Stochastic Machine

So far, for the sake of simplicity, I have limited my discourse to changes in the inanimate world. But now I have to turn to the biological world because, after all, it is here where free will is most relevant. All animals interact with their environments, made possible by the function of the brain, which follows physical laws like any other piece of matter. Further, the brain is an enormously complex entity, comprising 100 billion neurons (rivaling the number of stars in our galaxy) and no fewer than 100 trillion synaptic connections. Like Boltzmann's bag of gas molecules, but even more so, the behavioral outcome of this complex system is inherently unpredictable. Being a *chaotic* system, the future events taking place in the brain are as hard to foresee as the weather conditions ten days down the road.[11] Besides, the brain can be in a *metastable* state where the next step is unpredictable. We see this in a dog under provocation, oscillating between fear and aggression, leading to either retreat or attack, or a turkey being chased for dinner, veering randomly left or right.[7] At the human level, decisions made in everyday life frequently involves metastable dynamics, wavering between opposite tendencies with the final outcome landing on one side. Therefore, taken as a physical entity, even mathematically deterministic changes in the brain cannot guarantee a behavioral outcome that is always predictable to an observer.[12,13]

However, more importantly for the topic of free will, other than what is described above, the brain has an additional layer of complexity not found in any other physical entity, namely, the issue of the mind. The question is, does mind contribute to the unpredictability of behavior? Consider the following scenario. I go to lunch every day at noontime. Knowing this, my friend one day predicts that I will go to lunch at

12 o'clock. But knowing that my friend makes this prediction, and trying to defy it, I respond to the prediction by not going to lunch that day. Knowing that I will defy him, on another day my friend announces his prediction that I will go to lunch, but secretly predicts that I will not. But this time I purposely go along with his public announcement and go to lunch at noon. My friend is baffled. To an observer, my behavior is unpredictable, yet each time I do it with full deliberation and do not let chance "throw dice" inside my brain. One way to explain this dilemma is to look at the brain as a convergent point of multiple causes, and to assign a selective role to the brain, which at any given time allows a certain cause but not others to go through and exert an effect. It is like a light filter that permits light of a certain wavelength to pass but blocks all others. The light that passes follows natural laws, but the selective function of the filter is what makes the difference.

Here lies the crux of the problem. If we insist that the selector is still caused by physical events, then it is just an extension of the physical causal chain (more of the same). Conversely, if we postulate that the selector involves something other than ordinary matter (i.e., the mind), then the argument takes a different turn. There are two possibilities: (1) the brain is stochastic because the physical causes converging on the brain are complex; (2) the brain is stochastic because mind intervenes with the physical events of the brain. Note that the two possibilities are not mutually exclusive. That the first is true is beyond doubt; whether the second is real is controversial and depends on where we stand on the mind-body issue. The answer would be simple if we assume that mind does not exist, for in this case all that happens in the brain could ultimately be explained in physical terms, and free will would just be an illusory (or self-deceiving) psychological phenomenon, despite the sense of agency it provides. In this scenario the difficulty in predicting animal behavior could be simply compared to guessing when the next tornado would strike.[14] If, on the other hand, we assume that mind does exist and is distinct from the physical brain, we cannot rule out its possible influence on the physical goings-on in the brain. Please note that this statement should not be taken to mean that mind cannot be caused. It only means that causation involving mind is of a different nature than simple physical interactions. Therefore, the ultimate answer to the problem of free will rests on whether a participating mind coexists with the brain, a point to be taken up in *Chapter 13: Self from Within: The Introspective Self*.

11.6 Can Free Will Alter the Course of a Chaotic World?

Sure enough, free will does make a difference to the world, to a limited extent. If free will is biologically adaptive, it has to alter the environment for the benefit of the *self*. It does this by *changing the probability of events, making them more likely, or less likely, to happen.*[15] When an event approaches certainty, free will is futile. Free will is most effective when the event is in the mid-range of the probability scale (0.50). Although I cannot change the whole world, I can change a small part of it, under certain conditions. For instance, I can duck a falling rock to save my life when I see it rolling down my way, but I cannot deviate the course of an asteroid hitting the Earth. That would need the combined free will of many, many people, using technologies of an enormous scale. On the other hand, even the combined technologies of the entire human civilization would be useless to avert the impending collision of our Milky Way and the Andromeda Galaxy, scheduled to take place four billion years hence.

11.7 Free Will in Lower Animals?

If we take free will as a biological phenomenon, it is not hard to assume that all animals have a will, in a degree proportional to their evolutionary status. Pet owners, especially of dogs, cats, and horses, know their individuality and respect their choices and preferences. Bees and ants, on the other hand, are primarily instinctive in that they are less capable of thinking, if at all. But within a limited range they, too, seem to have some flexibility of individual choice. Nigel Franks of the University of Bristol

Fig. 11.1. Aerial view of Coney Island Amusement Park (New York) on a Sunday afternoon, showing humans dispersed on the beach, an outcome of the attractive forces from such desirable stimulants as food, sex and entertainment, and the repulsive force to maintain minimal body distance. A scene like this can easily be compared to a swarm of ants. [Courtesy Rockefeller Univ. Press.]

studied the behavior of individual ants and observed that they demonstrate teaching and decision-making, even quick decisions in times of crisis.[16] Fruit flies also behave in a spontaneous, individually distinct manner. Humans, to be sure, enjoy individual freedom and deliberation, as we intuitively know. But when observed *en masse*, humans may appear not much different from a colony of bees or ants (Fig. 11.1).

I like to accord lower animals some sort of free will, as they also have goals to attain, the main difference being that their goals are not as long-range and complex as ours.

11.8 Neurophysiological Basis of Volitional Acts

As shown in Figures 11.2 and 11.3, decision-making comes in several stages and involves multiple factors. Once a decision is made in the prefrontal cortex, a chain of command is transmitted sequentially to the supplementary motor area (SMA), the premotor area, and the primary motor area. From the latter the descending nerve impulse is transmitted to the musculoskeletal system to initiate a volitional act. (The corresponding brain areas are shown in Fig. 11.4.) The resulting movement

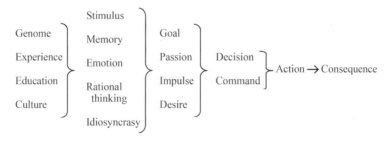

Fig. 11.2. Flow chart of decision-making. The execution of a voluntary action proceeds in stages. First, the background consists of the innate and acquired make-up of the agent. The dynamics of decision-making are the interplay of the external stimulus with memory, emotion, rational thinking, and idiosyncrasy. Idiosyncrasy confers individual bias at a given instant, making free will somewhat unpredictable and irreproducible. I use the term "idiosyncrasy" broadly to accommodate all the contingencies that can arise in the brain out of complex interactions, including incidental variables, deterministic chaos, catastrophic phenomenon, and other metastable conditions. A very small bias at the moment can tip the balance of decision one way or the other. From this dynamics, a goal is set and a drive is generated to achieve it. Finally, a command is sent from the motor areas to the skeletal muscles to start an action. (Please note that in humans a decision need not be immediately translated into action.)

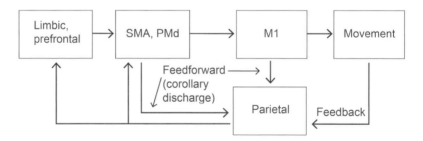

Fig. 11.3. Diagram showing neuroanatomical pathways of volition. Starting from the left, a decision is initiated in the prefrontal cortex upon interaction with the limbic system. The decision is translated into an action through the motor areas of the cerebral cortex. SMA, supplementary motor area; PMd, pre-motor area (dorsal part); M1, primary motor area. Note that the sense of agency of an act is achieved when feedback information matches the feed-forward in the parietal cortex. [See Note 30; permission Springer Berlin.]

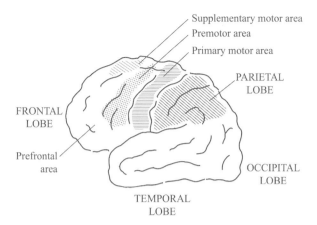

Fig. 11.4. Human brain (left cerebral hemisphere viewed from the outside, head pointing left) showing areas of the cerebral cortex involved in volitional action. [Refer to Fig. 11.3; also compare with Fig. 7.10 in *Chapter 7*.]

is fed back from muscle, tendon, and vision through sensory fibers to an evaluative center in the parietal cortex, and eventually to the supplementary motor, the premotor, and the prefrontal areas. The constant sensory feedback provides a mechanism for fine-tuning the motion. In addition, through a corollary discharge system, feed-forward signals are sent from the supplementary motor area, the premotor area, and primary motor area to the parietal cortex evaluative center, where comparison is constantly made with the feedback information. A correct *matching between feed-forward and feedback messages likely produces a sense of agency*, the subjective feeling of work accomplished (volition). It is interesting to note that electrical stimulation of the inferior part of the *posterior parietal cortex* alone induces the experience of intention to move as well as the experience of motion, even though no actual muscle movement has occurred, thus implicating the participation of the parietal cortex in the generation of "free will."[17]

I should hasten to add that, although the matching of forward and backward messages in the parietal cortex satisfies the sense of agency in terms of a single voluntary act (such as flexing the arm), it is far too simplistic and inadequate to explain the complex and subtle behaviors, such as deciding whom to vote for during a presidential election.

11.9 Free Will was Almost Scuttled

It will be tempting to put free will into experimental test. This will not be too hard if we can find an electrical activity in the brain that correlates with the subjective feeling of being a free agent of an action. In 1965, Kornhuber and Deecke reported that the performance of spontaneously generated voluntary actions were preceded by electrical changes in the brain, called *readiness potential* (RP), recorded on the scalp EEG (electro-encephalogram).[18] Subsequently, Benjamin Libet devised a method to time the subjective sense of volition and the appearance of readiness potential in reference to the time of muscle movement (recorded by electromyogram or EMG). The subjects were told to watch a clock and note the position of the second hand when they feel the urge to move — in this case flicking of the wrist. Contrary to expectation, the study showed that readiness potential preceded the action by 550 milliseconds, whereas the wish to act preceded the action by only 200 milliseconds. In other words, the brain signal to move took place before the intention to move.[19] This anti-intuitive finding casts doubt on the experience of agency, and has been interpreted by some to indicate that the subjective feeling of will is only an illusion, not the true cause of an action. It is as if the conscious brain is not the decider, but a reporter of the decision that has already been made unknowingly.[20]

However, Libet's discovery has alternative explanations. First and foremost, intention appears in different forms (plans, goals, purposes, decisions, choices, and so on) and in different stages. For instance, a college student opens a textbook of chemistry and turns to page 65. What are the intentions preceding this simple act? We may find out that he is interested in natural science, that he likes chemistry, that his ambition is to be a chemical engineer, that he enrolls at the University of Illinois, that he registers for a course in introductory chemistry, that there will be a midterm examination tomorrow, that he needs a good grade to pass the exam, that he has to know chapter 4 well, that he has to locate page 65 where chapter 4 is, and that he has to twist his right thumb along with the index finger to turn the pages. Which of this series of intentions is the cause of the action — turning the book to page 65? The proximate cause is the motion of the fingers, but this is only a minor part of a long series of goal-directed activities (see diagram in Fig. 11.2). Some of the intentions are remote, some are intermediate, while some are immediate (muscle movement). Libet and Wegner's interpretation[20] of intention and free will based on readiness potential could be far too narrow. Although a common endpoint of all volition, voluntary muscle contraction is just a part of the activity resulting from free will.[21] It has been suggested that in Libet's experiment, the effective intention might have started when the subjects agreed to participate in the experiment, long before the twitching of a muscle.[22]

Besides, Libet himself suggested that the agent could abort the action after feeling the urge to act, questioning readiness potential as the irreversible cause of an action.[23] This "veto power" that overrides the readiness potential signifies the ability of the brain to suppress an initial impulse and has been interpreted by some to be a free choice — or the "real" free will. A brain area called the dorsomedial prefrontal cortex has been found to be activated during this inhibitory function.[24]

In short, although the exact timing of volition is currently mired in controversy, the idea that the subjective sense of agency of an act derives from the matching of the feed-forward and feedback messages in the brain remains a plausible hypothesis (Fig. 11.3).

11.10 Pathological States of Volition

Alien hand syndrome refers to involuntary, complex motor behavior observed in patients with damage in the premotor and supplementary motor areas, leaving an intact primary motor cortex. The hand does not obey the command of the agent, as if it had "a will of its own." The hand can engage in mischievous behavior against other parts of the body, such as pinching the nipples.[25] Functional MRI study of a patient exhibiting alien hand syndrome revealed the activation of the primary motor cortex only, whereas voluntary hand movements of the same patient were correlated with activation of other areas as well, including the premotor and supplementary motor areas.[26] These cases illustrate that, in a pathological state, a person can be the causative agent of an action without the conscious feeling of being one.

The parietal lobe is responsible for dynamic representation of bodily image, including that of the limbs. Patients with parietal lobe damage may confuse their hand movements with those of other people. This corroborates the notion that the parietal lobe is involved in the sense of agency of an act.[27]

Frontal lobe lesions lead to symptoms related to loss of volition and executive function. *Abulia* refers to the absence of initiative and motivation, whereas in *akinetic mutism* (a more severe form of *abulia*) there is lack of spontaneous movement and the urge to communicate.

Illusion of voluntary limb movement can happen in people immediately after amputation, provided that the limb was functional before the event.[28] Such phantom movement of a lost limb depends on an intact parietal lobe.[27,29] Some schizophrenic patients, when moving their body parts, may have the delusion that such movements are under the control of someone else. These patients lose the sense of agency for their movement.

Hypnosis and brain washing can distort volition. If a person under hypnotic trance is instructed to perform a certain act, like opening a window, he may execute the order upon waking up, without knowing why he does it. If asked, he may rationalize the action with irrelevant explanations like the room is too hot, when in fact it is not. Long-term indoctrination (brain washing) has the same effect as hypnosis.

Notes and References

1. James W. (1890) *The Principles of Psychology.* Vol. 1, Henry Holt; Dover, New York, 1950, p. 453.

2. Thanks to Prof. Laird Addis who pointed out to me the futility of defining freedom objectively.
3. The affirmation of *self* through free will was strongly stressed by Sartre in his existential philosophy, when he said that a man "is what he wills." See: Sartre J-P: *Existentialism and Humanism*. Lecture given in 1945, published in French in 1946; translated into English by Mairet P, published in 1948 by Methuen & Co. London; quoted from p. 28 of the 1968 printing.
4. Here, the reader is reminded that, under the democratic principle, a person is free to act in so far as the act does not infringe on other persons' freedom, and that a certain degree of freedom is to be sacrificed in exchange for social cohesion.
5. Newtonian physics is applicable to objects of the scale commensurable with the human world. Objects smaller than an atom follow quantum mechanical principles, which are inherently unpredictable (in terms of position and momentum) because of their wave function. However, quantum uncertainty does not concern our everyday life.
6. Lorenz EN. (1963) Deterministic nonperiodic flow. *J Atmospheric Sci* **20:** 130–141.
7. Zeeman EC. (April 1976) Catastrophe theory. *Scientific Am* **234:** 65–83.
8. Dennett DC. (1984) *Elbow Room*. MIT Press/Bradford Books, Cambridge, MA.
9. Russell B. (1975) *The Problems of Philosophy*. Oxford Univ. Press, London, Chap. VI, On Induction. Some authors prefer a slightly modified version in which a turkey was fed starting from the day after Thanksgiving and was led to think that it would be fed forever until it was slaughtered on the next Thanksgiving Day.
10. The statistical principle of Boltzmann should not be confused with the uncertainty principle of quantum mechanics. Boltzmann's principle applies to molecules whose individual motions are predictable in isolation but become unpredictable *en masse*. By contrast, quantum uncertainty is the *inherent* unpredictability of matter at the subatomic scale.
11. Even a simple neural network is a nonlinear, vastly stochastic system that is vulnerable to deterministic chaos to a high degree.
12. Glimcher PW. (2003) Decisions, uncertainty, and the brain. MIT Press, Cambridge, MA.

13. Kelso JAS, Tognoli E. (2009) Toward a complementary neuroscience: Metastable coordination dynamics of the brain. In Murphy N, Ellis GFR, O'Connor T. eds.: *Downward Causation and the Neurobiology of Free Will.* Springer-Verlag, Berlin.
14. If mind does not exist, or if mind exists but does not affect the brain, free will is reduced to a series of physical processes playing out in the cerebral cortex. The scenario is thus similar to changing weather conditions, only more complex. Why, then, not accord free will to the weather, if the presence of mind is irrelevant? If we refuse to do so, we have to admit that there must be a qualitative difference between weather and animal behavior. And if so, what is the difference? What makes this difference possible? More discussion follows in *Chapter 13*.
15. This view is consistent with the probabilistic theory of causation. See: Suppes P. (1970) *A Probabilistic Theory of Causation.* North-Holland Publishing Co. Amsterdam.
16. Morell V. (2009) News focus: Nigel Franks profile. *Science* **323**: 1284–1285.
17. Desmurget M, Reilly KT, Richard N, *et al.* (2009) Movement intention after parietal cortex stimulation in humans. *Science* **324**: 811–813.
18. Kornhuber HH, Deecke L. (1965) Herpotentialunderungen bei willkurbewegungen und passiven, bewegungen des menschen: bereitschaftspotential und reafferente potentiale. *Pflugers Archiv Pysiologie* **284**: 1–17.
19. Libet B, Gleason CA, Wright EW, Pearl DK. (1983) Time of conscious intention to act in relation to cerebral activity (readiness-potential): The unconscious initiation of a freely voluntary act. *Brain* **106**: 623–642.
20. Wegner DM. (2002) *The Illusion of Conscious Will.* MIT Press, Cambridge, MA.
21. The outcome of muscle contraction has to be broadly interpreted to include writing (hand movement) and speech (vocal cord function).
22. Mele AR. (2009) *Effective Intentions: The Power of Conscious Will.* Oxford Univ. Press, Oxford; Mele AR. (2014) *Free: Why Science Hasn't Disproved Free Will.* Oxford Univ. Press, New York.
23. Libet B. (1999) Do we have free will? *Journal of Conscious Studies* **6**: 47–57; Libet B. (2004) *Mind Time: The Temporal Factor in Consciousness.* Harvard Univ. Press, Cambridge, MA.

24. Brass M, Haggard P. (2007) To do or not to do: The neural signature of self-control. *J Neurosci* **27:** 9141–9145.
25. Doody RS, Jankovic J. (1992) The alien hand and related signs. *J Neurol Neurosurg and Psychiatry* **55:** 806–810.
26. Assal F, Schwartz S, Vuilleumier P. (2007) Moving with or without will: Functional neural correlates of alien hand syndrome. *Ann of Neurol* **62:** 301–306.
27. Blakemore S-J. (2009) How we recognize our own actions. In Murphy N, Ellis GFR, O'Connor T. eds. *Downward Causation and the Neurobiology of Free Will.* Springer-Verlag, Berlin, pp. 145–151.
28. Ramachandran VS, Hirstein W. (1998) The perception of phantom limbs. *Brain* **121:** 1603–1630.
29. Frith CD, Blakemore SJ, Wolpert DM. (2000) Abnormalities in the awareness and control of action. *Philos TransR Soc Lond B* **355:** 1771–1778.
30. Hallett M. (2009) Physiology of volition. In Murphy N, Ellis GFR, O'Connor T. eds. *Downward Causation and the Neurobiology of Free Will.* Springer-Verlag, Berlin, pp. 127–143.

Chapter 12
The Expanded *Self*: Society as *Self*

Self is nested as a member of a larger self, which in turn becomes a member of an even larger self ...

Overview: *Individual selves can coalesce into a bigger self if the latter forms a tightly knit unit subordinating the former. Thus, a single cell existing alone is a self, yet an organism comprising millions of cells is also a self — now a mega-self. Unlike social insects that form mega-selves solely on the basis of instinct, human societies are much more complex and flexible. In forming a human society, social cohesion conflicts with individual freedom, and a compromise constantly needs to be made. A society can last as long as its members receive more than what they have to sacrifice. Humans have an inborn social conscience shaped by thousands, perhaps millions, of years of relentless group selective pressure in the course of evolution. Thus, in a successful society most members are moral — that is, pro-social — by definition, otherwise it would have collapsed and perished.*

Most wars result from, directly or indirectly, competition among human groups for limited resources. It is a competition among mega-selves for survival, a form of group selection. Groups that possess a high degree of internal cohesion are more likely to succeed. It is also a fact that excessive intra-group cohesion tends to enhance inter-group cruelty and atrocity. Thus, human nature can be at once good and bad depending on the vantage point — inside or outside a group. War will never end until all of humanity becomes a single mega-self.

12.1 The Stratification of *Self*

I define *self* as a system that functions for its own continued existence. Therefore, *self* can fit into any mold as long as the latter conforms to this scheme. *Selves* can be stratified hierarchically, in a manner that a member at one level enwraps those of the next lower level. We see this phenomenon in cells that form tissues and organs and finally the entire complex organism. In human society, individuals group into families, clans, tribes, and nations, each level preserving the characteristics of *self*. I call this stratified structure the expanded *self* or mega-*self*.[1]

When *selves* bond to form a mega-*self*, the following features are present: (1) a conflict of interest arises between the group and the individual members; (2) conflict resolution is needed to achieve harmony between the two, calling for compromise and constant adjustments (and readjustments) on the part of the individuals, in exchange for benefits derived from the group; (3) a successful mega-*self* is one in which the majority of individuals are better off with it than without it, even though some of the benefits they give up may appear to be self-defeating. For example, when groups of cells differentiate into a multi-cellular organism, the majority of them give up their prospects for immortality, but this is compensated by the perpetuation of the organism and its genome, resulting in more similar cells being formed as time goes on. In evolutionary sense, the individual members prevail in a mega-*self*.

12.2 Cooperation: Key to the Expanded *Self*

There is no denial that competition for limited resources is the business of life. These include food, water, foraging territory, farmland, and also space needed for shelter, breeding, and caring of the young. Living things adopt two opposite evolutionary strategies for competition. One strategy is to increase in number as fast as possible, as in bacteria. A typical bacterium can double every twenty minutes and, because of their rapid mutation rate and horizontal gene transfer, can come up with numerous genetic variations within a short time, guaranteed to survive in almost every possible and unexpected ecological niche. The second strategy is to increase the ability to alter the environment for one's own need, as seen in animals with intelligence and sophisticated body parts.

In this instance the organisms live longer but sacrifice their rate of procreation. Both strategies seem to work, as evidenced by their coexistence over eons of Earth history.

But there is a third strategy, which involves cooperation instead of competition. Cooperation takes place within species and also between species. Formation of an alliance (a form of expanded *self*) between disparate species provides a competitive edge over those outside the alliance, a reason why evolution did not allow a single species to drive all others into extinction.

Cooperation between species is called mutualism, whose extreme form is symbiosis. Mutualistic coexistence happens in our body, as we harbor trillions of *E. coli* bacteria (ten times outnumbering the total human cells) in our intestine, which suppress the growth of harmful bacteria and provide us with certain essential nutrients like vitamin K. Rumination of herbivores such as cows, sheep, and horses depends on mutualistic gut fauna that help them break down cellulose from plants they eat. Likewise, termites owe their wood-eating ability to microbes living in their gut. Coral reefs are formed as a result of mutualism between coral organisms and the photosynthetic algae that live inside them. Mutualism also exists between plants and the mycorrhizal fungi; the former fix carbon from the air, whereas the latter help in extracting minerals from the ground. Other types of mutualism include pollination of flowers by honeybees. Some plant species and insect pollinators are so interdependent that one will not survive without the other, resulting in co-evolution. Other plants expel harmful insects by attracting insect-eating carnivores. Even a simple relationship with the food we eat is an example of cooperation. Take apples for example. We cultivate apples to satisfy our taste and nutrition, but the apple trees entice us to help perpetuate their species by giving us tasty fruits.

Intra-species cooperation occurs all the time. This happens even among non-sibling members. A glaring example is seen in the greater anis (*Crotophaga major*), a neotropical cuckoo. These birds form monogamous pairs that are not immediate relatives. Groups of up to four pairs share a nest in which all the females lay their eggs. All members of the group participate in the defense of territory, incubation of the eggs, and feeding of the young. Adults cannot recognize their own eggs or nestlings so no preferential treatment is possible. In this communal breeding, individual fitness of all the pairs increases. Bigger pairs are able to defend nesting sites better than smaller pairs, and pairs attempting to breed alone invariably fail. This is a success story of non-kin cooperation within the same species.[2]

12.3 Lessons from Microbial Communities

Bacteria are commonly viewed to be individualistic and enjoy independent existence. More recent studies, however, revealed certain degrees of social interaction. Quorum sensing is a decision-making process used by bacteria to communicate to one another a change in environment, as a result of which they switch from a nomadic life to one in a biofilm, where they survive better *en masse* and become resistant to antibiotics. Bacteria of different species can easily undergo "horizontal" gene transfer through which they exchange their genetic materials to form new species. It is believed that such genetic recombination accounts for 80% of evolution of bacterial proteins. Different species of bacteria also form metabolic alliances, in that the end product of one becomes the nutrient of another, and *vice versa*.[3] Certain bacteria in the wild are organized into socially cohesive units in which antagonism occurs between, but not within, ecologically defined populations. Bacterial clusters known to have cohesive habitat association also act as units in terms of antibiotic production and resistance. Within a population, antibiotics produced by a few are tolerated by all other members, suggesting cooperation among the individuals.[4]

Yeast cells respond to their environment in a collective manner. It was reported that when food source is scarce, there is a selective pressure for yeast to grow in clumps. Yeast cells normally secret the enzyme invertase into the medium to split sucrose into glucose and fructose, products that the cells can take in and digest. But when the concentration of sucrose in the medium is low, forming an aggregate gives them a better chance to take up the split products than as single cells, simply because of proximity to the useful nutrient.[5]

Perhaps the most spectacular social behavior of single cells is seen in the slime mold *Dictyostelium discoidium*. The life cycle of these tiny amoeba alternates between single and communal existence, and provides the first evidence of division of labor — differentiation. (For details, see Fig. 4.3 in *Chapter 4: The Microbial Self.*) When nutrients are abundant, they behave as individuals; when food is scarce, they aggregate and turn into a slug, which literally "walks away" in search of "greener pasture." In the new environment the cells differentiate into a fruiting body consisting of a stalk and spores, each with a unique function: the spores are released and are capable of germinating into single cells for the continuation of the species; the stalk cells support the spores and help in their dispersion, while sacrificing their privilege of infinite propagation. Slime molds straddle the immortal unicellular life and the limited lifespan of a metazoan. The reproductive altruism of slime mold stalk cells serves as a prototype of differentiation in higher organisms.

Endosymbiosis of two distinct single cells is equivalent to merging two *selves* into one. Today, the most plausible explanation of the origins of certain organelles (the enclosed structures within a eukaryote) is the engulfment of one free-living organism by another. The plastids or chloroplasts of plant cells are believed to have come from cyanobacteria (bacteria capable of photosynthesis), whereas the mitochondria in all eukaryotes are believed to have descended from proteobacteria.[6] In fact, vestigial DNAs are still present in chloroplasts and mitochondria, betraying their foreign origin. Nonetheless, most genes in the endosymbionts today have long been transferred to, and integrated with, the host's nuclear genome, from where the protein products are allowed to go back to the organelles to carry out their native functions — photosynthesis in chloroplasts and oxidative phosphorylation in mitochondria. Historically, the functions added by the symbionts to the hosts enlarged the scope of environment in which the cells could survive, and provided considerable evolutionary advantage. (See Fig. 3.1 in *Chapter 3* for cell organelles.)

12.4 Lessons from Insect Communities

Intra-species cooperation is highly developed in insects. The most impressive prototypes are the social insects such as honeybees, ants and termites. These animals exhibit the phenomenon of eusocialism, a hierarchical, genetically determined social caste system, with strict division of labor including a reproductive role, that molds the entire community to behave like a super-organism, in which the constituent *selves* become subservient to the mega-*self*. The dedication of the individual insects to the colony is comparable to the dedication of the individual animal cells to the entire body.

Bees and ants both belong to the order Hymenoptera (having membranous wings). In the honeybee society, each hive is made up of a single queen, a few hundred drones, and up to 50,000 workers. The queen lives to about two years and lays about 1,500 eggs per day. The drones come from unfertilized eggs; they live to eight weeks and their sole function is to mate with the virgin queen. The great majority of a colony consists of workers, which are fertilized females. Only one among the fertilized females becomes queen, after being fed with royal jelly when young. The reproductive function of the workers are suppressed by a hormone produced by the current queen, but in the event the queen dies, one of the workers will be chosen to be the new queen. As the name implies, worker bees tend to all kinds of chores including nursing the young, housekeeping, construction and defense of the hive, and foraging. Workers have a short lifespan, only six weeks at the most, having literally "worked themselves to death."

Like honeybees, ants are highly eusocial. The main difference is, while all species of honeybees engage in the same kind of business (honey collection), ants evolved into multiple-task species, each species doing a specific work distinct from others, and each depending on one type of food for subsistence. Today there are fourteen thousand known species of ants. Their high degree of cooperation and evolutionary adaptability makes them the most successful animals on earth, surpassing all other insects. Their total biomass occupies more than half that of all insects combined, and is roughly equivalent to that of all mankind.

The social structure and genetic features of ants are roughly the same as those of honeybees. Namely, a colony is made up of one (or several) queen(s) whose only function is to lay eggs and reproduce; a small number of males whose role is to fertilize the queen; and numerous workers (all females) that engage in caring for the queen and the brood, in maintaining the nest, and in food gathering. Some workers double as soldiers for defense and attack. Among the countless species of ants, a few are especially interesting and worth mentioning.

Leafcutter ants are peaceful farmers and vegetarians. Their workers cut leaves into fragments and transport them to the nest to construct a garden, in which they grow fungi for food consumption. By contrast, army ants (and the African driver ants) are an aggressive group whose voracious, insatiable, carnivorous appetite drives them on perennial migration in search of food. Their long list of prey consists of other insects including roaches, beetles, wasps, and other ants; and even small animals such as scorpions, rodents, snakes, lizards, and birds. A marching swarm of army ants is one of the most unforgettable displays of nature, fanning out several miles long and hundreds of feet in width. In times of danger, the workers use their bodies, frequently at their own peril, to protect their charge, and build living bridges with their bodies to allow other members to march on. These ants are active in the day time but rest in temporary camps at night in a "bivouac", made up of the swarm itself. To do this, between 150,000 and 700,000 workers link their bodies and legs together with their claws, forming interlocking layers of chains and nets and solid walls that safely surround the queen and the immature ones. Ahead of their advance, army ants frequently send out scouts that leave a scent trail for the colony to follow. Knowing this, natives of Africa utilize the ants as a natural pest control. Upon seeing the scout ants, the natives will vacate the village and return a few days later, whereupon their premises will be free of spiders, snakes, rats and other noxious pests, all eaten by the ants. Direct encounter with the army, of course, is life-threatening to people.

Slave-maker ants are a different breed. They make their living by capturing the worker caste from other colonies to work for them. In extreme examples, a slave-maker queen produces only soldiers, whose only task is to enslave foreign workers to feed the colony. These soldiers are fierce fighters, using their strong mandibles to tear the enemies in a gruesome battle. Typically, the queen sends out raiding parties to attack other ant colonies. The raiding parties slaughter the adults and carry the pupae back to the home queen, who then raises the adopted children as her new slaves to work for her colony. Another tactic used is for a new slave-maker queen to sneak into the invaded colony during a confusing fight and replace the enemy queen by killing her. The new queen mimics the old queen by consuming the latter and acquiring her scent (pheromones) to attract and subdue the attending ants. (Some invading queens will feign death at the entrance of a foreign nest and let the host workers carry her to the queen's chamber, where the invading queen will spring to life and murder the resident queen.) Subsequently the new queen (invader) produces more soldiers of her own, which are again sent out to raid other colonies to capture more slaves. It is interesting to note that in a slave-maker colony the enslaved workers care for the eggs and young that are not their genetic relative, often not even their own species.

Termites are also enormously successful eusocial insects. But please note that although they appear like ants, the two are genetically and evolutionarily very distant. Termites belong to the order *Blattodea*, which includes cockroaches. Like ants and honeybees, termites conform to a strict caste system with division of labor including reproduction. Termites typically live in the soil and consume cellulose from dead plants and wood. They are able to thrive on wood fibers by harboring cellulose-digesting microbes in their guts. Their food habit leads to their destructive power on buildings. However, in the wild, termites contribute to ecological balance by recycling wood and other dead plants. In an open space, when a colony gets big, the subterranean nest can grow above ground to form mounds of a few meters high, housing several million individuals. A colony is made up of nymphs (semi-mature young), workers, soldiers, and reproductive members of both genders, sometimes including several egg-laying queens, each capable of generating 2,000 eggs a day. The pheromone produced by the primary queen is believed to be the integrating element of the colony, which is spread by shared feeding. A queen can live to forty-five years and continue to mate with the king for life. The workers are the main caste; they digest wood and feed the other members of the colony. Termites are not predatory. They use soldiers only to defend against attacks from ants.

12.5 What Causes Insect Eusociality?

Haldane first proposed in 1955 that altruism in social insects has a genetic basis, that an individual is more likely to sacrifice himself to save the life of a close relative than an unrelated person simply because of genetic relatedness, and that altruism is just another way to promote one's own DNA. In 1964, Hamilton put this concept into a mathematical equation: $rB>C$, where B is the benefit and C the cost of an action, and r being the relatedness between the giver and receiver.[7] This model works nicely to explain the evolution of eusociality when considering the fact that *Hymenoptera* (to which both bees and ants belong) have a genetic peculiarity in being "haplodiploids."[8] Hamilton pointed out that in these insects, a sterile worker (female) is more related to her siblings (including the future queen) she cares for than to her own daughter if she were to reproduce, the relatedness being ¾ and ½, respectively, a situation not seen in diploid animals, where the values are ½ and ½ (see Hamilton W. D.'s article for details).[7] Therefore, according to Hamilton, that the worker bees and ants sacrifice their own reproductive chance in favor of their siblings actually increases selective fitness for the workers' own genes. Hamilton's idea is called "kin selection theory" or "inclusive fitness theory."

Hamilton's equation was taken as the rule for many years, until the genetic makeup of termites came to light in the 1990s. It was found unexpectedly that termites, which are incredibly successful eusocial insects like bees and ants, are actually *diploids* rather than *haplodiploids*. In other words, all members of a termite colony, male and female, are uniformly diploids, so the relatedness of a sterile worker to a sibling and to a daughter are no different (both being ½). Furthermore, it is now known that there are many haplodiploid insects that do not live in a eusocial structure. Hamilton's idea became problematic — haplodiploidy is neither a necessary nor sufficient condition for eusociality, though it may be a contributing factor. Recently, D.S. Wilson and E.O. Wilson pointed out the inadequacy of Hamilton's theory, which is based on selection at exclusively *individual* level, and proposed to replace it with a dynamic mechanism based on selection at the *group* as well as individual levels — called *multilevel selection*.[9] In short, the success of social insects is the outcome of a combination of group and individual selections. Altruism plays an important role in group selection as well as kin selection. As Wilson and Wilson suggested, within the same group, selfish members survive better than the altruists; but in a competition between groups, those with altruist members fare better than those without.[10,11] Interestingly, this is essentially the revival of Darwin's original idea laid down in *The Descent of Man* in 1871.

In learning the lesson from social insects, we must not forget that their group behavior is stereotyped and genetically programmed, with very little individual variation. They rely on group rather than individual identity. Their social determinants are mainly odors (pheromones), carried by a group of volatile, small hydrocarbon molecules produced endogenously or acquired from the environment (over twenty of these are known for ants). An odor, or an orchestration of odors, defines a caste status and nest identity. When two ants come in contact, the odors they "wear" can tell a friend from a foe, and the outcome is a matter of life or death.

By contrast, the degree of flexibility in the group behavior of humans is enormous. It would be a grave mistake to literally extrapolate social order from insects to humans. Unlike social insects, human groups can be formed or dismantled in the blink of an eye, based on changes in the self-interests of the constituents. We see this frequently among nations. For example, when World War II was raging, the United States and Japan fought tooth and nail. But as soon as the war was over, a new alliance was forged between the two almost overnight against the threat of the Soviet Union. On a personal level, an individual can at once be a member of several social groups — the same man can play the role of a loving husband, a stern father, a strict law enforcer, a friendly waiter, or a congenial sales person, depending on the time of day or the uniform he dons.

12.6 Forces Enhancing Social Cohesion

In order to reap the benefits of a group, humans need to partially suppress their idiosyncrasies and the tendency to seek individual freedom. Thus, in a society, we are both submissive and rebellious. Social cohesion and individual liberty are the antitheses that have to be adjusted and balanced in order to arrive at a harmonious society.

What holds the *selves* together is what I call "social glue." The societies of honeybees, ants and termites make use of the most primitive of "glues" — the genes and chemical messages (pheromones). The bonding forces in human societies are much more complex, covering the entire gamut of biological and cultural factors, from genetics to physiology, biochemistry, psychology, education, social custom, mythology, religion, and all of civilization. The important factors are reviewed below. Note that some of these factors apply to lower animals as well.

12.6.1 *Genetic factors: Instinct and conscience*

As discussed above, cooperation among close relatives is only part of the story. In a heterogeneous society, humans have an instinctive propensity to help out others in need even if they are not genetically related. This inborn *conscience* is engrained in the genome of humankind, "sculpted" by millions of years of relentless natural selection. The reason is, a species whose members help one another stands a better chance of survival, and therefore is favored by evolution. Human conscience constitutes a strong social cohesive force.

12.6.2 Basic physiologic needs

Smell and touch are primitive factors involved in social bonding. Odor can evoke a strong emotional response, and a person may feel attracted to, or repulsed by, another because of body odor, sometimes even below the threshold of awareness. Odor preferences may be inborn or learned at a young age. For example, a tempestuous toddler can easily be calmed down by the familiar smell of a security blanket. People may become homesick if they miss the unique odor of their old houses. And think of the billions of dollars men and women spend each year on perfumes. Touch is another socially relevant stimulus. Hugging, caressing, kissing, and intercourse are different forms of tactile sensory satisfaction animals and humans seek. For a newborn, the texture and warmth of a mother's breasts provide one of the earliest bonds between mother and child. The softness of a stuffed animal provides security to most infants.[12]

12.6.3 Pro-social hormones

The effects of estrogen and androgen on sexual attraction and virility are well known. Among other hormones, two stand out as particularly relevant to social bonding: oxytocin and vasopressin, both small peptides endogenous to the brain.[13] In the nervous system, oxytocin and vasopressin have potent neuro-modulatory effects through activation of the dopamine motivation/reward system, in particular the nucleus accumbens and ventral pallidum of the forebrain. (See *Chapter 9: Self and Emotion* for the brain structures.) Strong effects of these substances on pair bonding have been demonstrated in field rodents.[14] For example, *prairie voles* exhibit a monogamous social structure while closely related *meadow voles* are polygamous. Pair bonding behavior in prairie voles is correlated with a high density of vasopressin receptors in the ventral pallidum, and of oxytocin receptors in the nucleus accumbens. Bonding in these animals is lost when these receptors are blocked.[15] On the contrary, male meadow voles, which normally are promiscuous and do not form male-female bonds, can be induced to express pair bonding by introducing vasopressin receptors to the ventral pallidum.[16] In another experiment, female mice whose oxytocin receptor gene has been inactivated show a deficit in maternal behavior.[17]

Pro-social behaviors of oxytocin in humans are well recognized. After childbirth, oxytocin motivates the mother to nuzzle and protect the newborn, forging a strong mother-child relationship. In both sexes, oxytocin induces care, warmth, tenderness, attachment, cooperation, trust, generosity, and empathy.[18] However, when humans are divided into groups, it was found that oxytocin motivates in-group conformity and favoritism but promotes out-group derogation, implying hormonal influence on ethnic prejudice and xenophobia.[19] Another example of neuro-hormonal influence on social behavior is seen in desert locusts. These insects live a solitary life when population density is low. But when the population reaches a critical point, they start to aggregate and swarm. The switch from initial mutual repulsion to strong attraction is triggered by the neurotransmitter serotonin.[20]

12.6.4 *Early life experience*

Humans and other animals are attached to things they experienced in early life, sometimes within a critical period. Perhaps the most dramatic example is the case of "imprinting" demonstrated by Konrad Lorenz. As goslings are attached to the first moving object they see after hatching, Lorenz was able to show that they preferred to follow him instead of their natural mother. Humans learn their native tongue easily and become endeared with it, while second and third languages are never as intimate and personal. Most people acquire their religious preference from their parents during childhood. Westerners who are accustomed to diatonic music may dislike the pentatonic melodies of the East, and *vice versa*. In general, people who grow up in similar physical and social environments develop closer ties with one another.[21] Furthermore, those who have gone through the same hardship (at any age) such as war or natural calamities often forge a lifelong comradeship.

12.6.5 *Group pressure*

Social conformity can be the outcome of peer pressure. The following classical experiment, performed in the 1950s, is revealing. A group of college students were asked to compare the length or height of a series of lines (see Fig. 12.1). While doing so, each student was permitted to observe the results of others. But unbeknownst to one of them (the test subject), the results given by others were fake and intentionally incorrect in a consistent way. The surprising outcome was that, in most instances, the test subject preferred to blindly follow the judgment of others (the majority), even if the judgment was initially considered wrong by him.[22] It thus appears safer for an individual to follow the group than to formulate his own opinion.

12.6.6 *Collective consciousness*

Members of a society usually share a common cognitive content — experiences and memories in the form of oral and written histories, which may include hard-to-verify happenings in the remote past such as legends, myths, and stories of cosmic and human origins. Values, ways of life, aspirations, behavioral norms, existential meanings, and interpretations of death and afterlife (such as those contained in a religious belief) are part of this spectrum. These cognitive contents permeate the entire community and acquire a universal character among its members, are

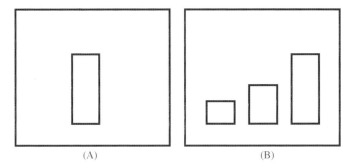

Fig. 12.1. Group conformity test similar to the one designed by Solomon Asch. The test subject is asked to match the height of the bar in the left box with one of the bars in the right box. Most people will yield to group pressure and make a wrong choice even if they initially have the right answer. [See Note 22.]

handed from generation to generation (instilled during the formative years) with little modification, and are distinct from private consciousness unique to each person. It is not so much the validity of the content as the deep-rooted nature of the beliefs that makes it a strong social binding force.

12.6.7 *Collective sentiment and the social function of art*

In *Chapter 9: Self and Emotion*, I discussed art as an extracorporeal expression of emotion. Here, in the context of society, I propose that art serves as an important social glue. Comparative anthropology affirms the universality of art in human societies, regardless of the level of civilization. Music, dance, sculpture, painting, and religious rites are found in all human tribes and races. I see no difference between a stirring tribal war dance and a spirited college football fight song. And look at the festivities around the world — the Oktoberfests in the Bavarian mountains, the fiestas of Latin American villages, and the powwows of Native American tribes. Are they not channels for emotional outlet? True, art does not fill the stomach of an artist, does not provide shelter in bad weather, but it coalesces individual emotions into public feelings and sentiments, propelling a group to act in unison.

Perhaps I should mention in passing that some evolutionary biologists maintain an opposite view of art. They suggest that art, along with other aspects of human culture, are evolutionary "spandrels". (A spandrel is an unintended space in architecture, fortuitously created by some essential structures of a building, but itself serving no useful function.) According to Gould and Lewontin, who coined the term, a "spandrel" in biology is a non-adaptive characteristic that arises as a *byproduct* of other adaptive features, even though it may subsequently enhance the fitness of the species.[23] In my opinion, these people fail to recognize the importance of group selection and the role of art in the process.

12.6.8 *Empathy*

Empathy is the sharing by a person of the feelings of others — their joy, their sorrow, and their pains. One prerequisite is the ability to "read" the facial expression of other persons. Studies on human subjects revealed that facial emotional recognition involves several brain areas: the sensory cortices, the facial recognition area, the emotional centers, and the hippocampus.[24] Using functional MRI, Singer and coworkers localized the substrate for empathy of pain experience to the brain network associated with affective qualities, including the anterior insula and the anterior cingulated cortex, but not the parts associated with sensory qualities.[25] Furthermore, if empathy turns into compassion toward humanity (a higher state of feeling), the reward system including the nucleus accumbens is activated.[26]

The neural correlates of empathy are further advanced by the discovery of a group of neurons that fires when a subject (monkey) is watching another subject performing a certain act. The neurons that fire in the observer's brain are those that would have been activated if the observer were intending to execute the same action. These neurons, called "mirror neurons," appear to play a role in what we understand as empathy.[27] For example, a mirror neuron that fires when the monkey rips a piece of paper would also fire when the monkey sees or hears another monkey ripping paper. Different mirror neurons respond to different tasks, including such acts as putting peanuts into the mouth of the experimenter. Using single cell recording in monkeys, such neurons have been detected in the inferior premotor cortex and the inferior parietal cortex. Similar neurons have been detected in the human brain.[28] Although mirror neurons may not represent the whole story of empathy, there is little doubt that they are involved in social interaction.

Emotional contagion is the mimicking of the emotional state of one person by another, frequently occurring in a closely-knit group. When one person yawns, others also yawn. When one baby cries, all babies cry. It has been demonstrated that mice become "upset" when cage-mates are in pain. Mice show enhanced pain sensitivity when they see a familiar mouse in distress but not when the other mouse is a stranger.[29] Empathy also occurs in birds. In the highly monogamous zebra finches, females show signs of distress (as measured by their blood glucocorticoid level, a stress hormone) when they hear their male mates putting out a distress call, but not when they listen to calls from unknown males.[30]

There is evidence that rats are able to help one another in distress out of pure empathy. When a free rat was placed in an arena along with another rat (a cage-mate) trapped in a restrainer, the free rat would intentionally open the door of the restrainer to release the trapped rat. Rats would not open the door if there was no rat inside, or if the restrainer contained other objects. This happened in the absence of an apparent reward.[31]

12.6.9 *Altruism*

Altruism is the sacrifice of *self* for the benefit of a group (a mega-*self*). On the surface, altruism appears counter to natural selection, because it gives the selfish individuals a better chance to pass on their selfish genes to the offspring. But in practice, only a fraction of the altruistic individuals actually sacrifice their lives, leaving the surviving altruists a chance to pass on the gene. Take martyrdom (the extreme case of altruism) for example. If in a society 80% of the members have the potential for martyrdom, it does not follow that only the remaining 20% (selfish members) will be the eventual survivors. In reality it does not take all the altruistic members to die for the cause in order for the group to be saved (most likely 10% will suffice). Thus, in this instance there is still a 7 to 2 ratio (70/20) in favor of the altruistic gene. Applying this scenario to the small animal that emits a warning call in times of danger, only the first caller (one of the altruists) will be captured by the predator; all other potential altruists have a chance to hide and survive. Or take the example of the legendary Dutch boy who saved a community from flooding by plugging the leak on the dyke; the sacrifice of one altruistic individual increases the survival chance of all others (altruists as well as selfish members). In the end, natural selection favors groups with altruistic traits. (See Section 12.5 on *Eusociality*.)

Inside our bodies, all somatic cells (body cells that do not engage in procreation) are altruistic in the sense that they sacrifice their prospect for immortality in favor of the germ cells (ova and sperms). Altruism is dramatically played out in our blood stream, where the neutrophils (a type of white blood cells that serve as foot soldiers) literally explode themselves in the act of battling invading bacteria. In the bee society, stinging is a suicidal act to save the nest. During migration of driver ants, workers use their bodies to build bridges for the colony to move over obstacles, even facing certain death. Whereas altruistic behaviors in lower animals are largely innate and genetically programmed, self-sacrifice in humans is strongly influenced by culture.

The following true story illustrates a moving example of human altruism in the face of disaster. It happened during the recent tsunami in Japan. On 11 March 2011, a strong earthquake, Richter scale 9.0, struck the northeast cost of Japan, setting off a tsunami minutes later, severely damaging a nuclear power plant in the city of Fukushima. The cooling system of the plant was destroyed, leading to the meltdown of two of the six reactors on 15 March releasing radioactive material with a radiation level 800 times the recommended exposure limit for the workers. However, workers were needed to continuously cool down the overheated fuel rods in the reactors with seawater as an emergency measure. An agonizing dilemma confronted the authorities: to keep sending workers into an increasingly contaminated area in a last-ditch effort to cover the nuclear fuel with water, thus exposing the workers to surely lethal levels of radiation, or to protect the workers from radiation but risk further meltdown, leading to much larger release of radioactive material and subsequently endangering a much larger population. Under the circumstances, 750 workers were evacuated, but, altruistically, 50 people volunteered to stay behind, heroically sacrificing their lives for the common good. Without remonstration, they took it as part of their duty, facing the prospect of death to fulfill their obligation, saying, "This is our mission: protecting people with our lives." Their families accepted the decision without complaint.[32]

12.6.10 *Role of culture*

The role of culture in unifying a group of people and shaping their identity is indisputable. Those who speak the same language, practice the same customs, uphold the same values, and share the same religious belief are likely to stay together. Conversely, animosities and clashes frequently happen among peoples of different cultures, resulting in inter-group conflict.

Culture is the cumulative product of human activities transmitted across generations through imitation and learning. In rudimentary forms, "culture" is also found in non-human higher animals. For examples, bird songs acquire regional variations as in human local dialects and accents. Macaque monkeys on Koshima Island (Japan) learn the tradition of washing sweet potatoes before eating. African chimpanzees use sticks to "fish" out termites for food, and use stones to crack open nuts, both transmitted through teaching. Humpback whales learn a special way to procure food through sound communication.[33] Nonetheless, it is only in humankind that culture becomes a formidable social driving force, for better or for worse.

Human culture originates from the ability of the brain to manipulate symbols, defined as an object, an act or an idea capable of representing something other than itself, usually of an abstract and complex nature, as exemplified by language. A prerequisite for symbol formation is the ability to perform abstract thinking. Because of this abstraction, a symbol can carry an enormous amount of information. Think of the history of a nation rolled into a simple flag, or consumer expectation embedded in a company logo.

The relationship between gene and culture is part of the age-old debate of nature versus nurture. The truth is that they have parallel yet interactive developments. Culture has certain contingent determinants that impart local flavor, but culture also has a genetic basis that confers certain universalities. Thus, we all have an inborn sense of rhythm and melody, but these alone do not predict whether the music of a people will be the enlivened *contrapuntal* melody of the Baroque period, or the tranquil mood of an Indian morning *raga*.

Culture affects genome in a way more than most of us realize. One striking illustration of gene-culture co-evolution is the interaction of cattle domestication and the gene for lactose digestion called lactase. Lactose is the sugar component of milk. All newborn mammals have the ability to digest lactose because they have lactase in the intestine, but the enzyme level declines after weaning. Adult humans who do not have this enzyme are lactose intolerant — they have bad reactions to fresh cow milk and may have symptoms of abdominal cramps and diarrhea. Those adults who carry the lactase gene can produce the enzyme and have no problem drinking milk. It is interesting to note that the appearance of the adult lactase gene correlates with the domestication of cattle, which occurred six thousand years ago. Further, the gene is now most prevalent in the Scandinavian countries where over 90% of the population are positive. It happens that, in addition to being an energy source, lactose aids in calcium absorption, a process needed for maintaining bone integrity, important in the northern part of the world where sunlight exposure is deficient (sunlight stimulates the formation of active vitamin D in the skin, a factor needed for calcium absorption). Apparently, in the northern hemisphere where sunlight is deficient, people with the lactase gene are able to enjoy the benefit of cow milk. They live healthier and longer lives and have a better chance for procreation. Here, a gene has been selected by the culture of animal domestication.[34]

Another instance of culture affecting the genome is the case of sickle cell anemia. Sickle cell anemia is a disease when a person carries two copies (homozygous) of the mutated gene. But when a person has only one copy (heterozygous), he not only does not manifest sickle cell disease (a carrier) but also is resistant to malarial infection. Therefore, in a place where malaria is endemic, a carrier of the sickle cell gene lives longer. In Africa, malaria is rampant in places where mosquito breeding grounds are abundant in the form of stagnant water ponds, which frequently are an outcome of excessive deforestation. Therefore, a human practice (cutting down trees) becomes a selective force for the perpetuation of a gene (the sickle cell gene). Other examples include the alteration of human gut size since the invention of fire and cooking two million years ago,[35] and the higher alcohol tolerance of Europeans relative to Asians following greater alcohol consumption in Europe.[36] In short, the controversy over nature versus nurture is resolved if we consider evolution as multidimensional, that it is not only a matter of genetics but also *epigenetics*, the latter encompassing culture as one of the determinants.[37,38]

12.6.11 *Common external threats (real or imaginary)*

Social cohesion can be fostered strongly, sometimes rapidly, in the face of an external threat to a society's existence, such as an attack by a foreign nation. History shows that crafty rulers frequently fabricate outside threats as a ploy to restore their failed governing power. One of the most glaring examples of unity against a common threat was the engineering of the first crusade by Pope Urban II in 1095. In a most rousing and enticing speech against a backdrop of internal fractionation and disunity, the Pope called for a war against Muslims to recapture Jerusalem, the Holy Land, promising absolution and remission of sins for Christians who joined the cause, resulting in consolidation of his own weakening power at home.

12.7 Group Selection Revisited: *Self* as the Fundamental Unit

It is highly relevant at this point to take up again the topic of "What is the most appropriate level of natural selection?" Is it the gene, the individual, the kin, the clan, or is it multilevel? Richard Dawkins declared his position in his famous book, *The Selfish Gene*: "The muddle in human ethics over the level at which altruism is desirable — family, nation, race, species, or all living things — is mirrored by a parallel muddle in biology over the level at which altruism is to be expected according to the theory of evolution... I shall argue that the fundamental unit of selection, and therefore of self-interest, is not the species, nor the group, nor even, strictly, the individual. It is the gene, the unit of heredity."[39] Obviously, Dawkins puts the gene on the center stage, arguing that it is the "fundamental unit of selection." From this perspective he accepts kin selection, including altruism among close relatives, and rejects selection in larger groups, because only the former is "genetically relevant." (The other kind of altruism that Dawkins accepts is the reciprocal altruism in which the mutual advantage is immediately apparent.) Contrary to this, Wilson and Wilson effectively argue that natural selection occurs at multiple levels, from individual to groups of different sizes.[10] While I concur with the latter idea, I would like to propose the use of "self," with its implied expandability, as the appropriate unit for the multilevel nature of selection. An individual is a *self*, a tribe is a *self*, and a nation is also a *self*, as they all are systems that auto-perpetuate. Not only that, a university, a bureaucracy, a football league, and an international business conglomerate, are all *selves*. Each competes with other *selves* of the same nature and the same rank for "survival." Thus, using *self* as a unifying concept, the confusion regarding the level of natural selection is resolved. (See related discussion on natural selection in *Chapter 3: Self and the Beginning of Life*.)[40]

12.8 Social Behavior in Non-human Mammals

Before discussing human societies, for comparative purposes it would be helpful to have a glimpse of the social orders of some non-human mammals. Animals that live in tightly knit groups are favored by natural selection as they can achieve things that one can never do. One vivid example is a pack of African wild dogs overpowering a buffalo many times the body size of each. To form a working group certain order needs to be in place. Marc Bekoff, an ethnologist, observed that a society of wolves and coyotes are governed by strict rules. They have a rudimentary social sense, including acts of tolerance, trust, forgiveness, reciprocity and achievement of fairness among the members. According to Bekoff, these animals acquire their social code of conduct through ritual plays at a young age. For example, an invitation to play is displayed by lying down their forelimbs with their hind limbs standing — the "bow" posture (incidentally, we see the same gesture in domestic dogs). As a token of friendship, the stronger members will let the weaker ones "attack" them by rolling on the ground. If the exercise gets too rough, one party will "apologize" and display the bow posture again. Early exposure to these rituals educates the young to the acceptable behavioral norms. Any member of the pack who breaks the rules, such as taking more than a fair share of food, will be ostracized and disbanded by the rest, leading to its wandering away. Without the benefit of pack living, the excluded animal meets an early death. In experiments with rats, it has been shown that they will not eat food if doing so inflicts pain on other rats. It was estimated that among animals that live in groups, over 95% of their behavior is pro-social.[41]

12.9 Evolution of Human Societies

It is hard to trace back to the beginnings of human society, but the following picture emerges from archeological evidence and observations on extant hunter-gatherers. Primitive humans followed a set of strict social rules, the most severe being the taboos, violations of which met with severe punishment. Everyday behavior was shaped by customs and justified by mythology and legends, frequently associated with supernatural or invisible agents. Children emulated the behavior of their elders. Above all there was an inborn sense of conscience, which was spontaneous and intuitive. As societies grew in size, powerful rulers dominated the group and promulgated laws to be followed. Laws are based on prevailing moral norms but differ from the latter mainly in the clarity of content and in defining the degree of punishment for the violators.

Since the age of Enlightenment of the eighteenth century, the concept of *social contract* started to replace the system of absolute monarchy.[42] Social contract stresses that individuals give up some of their freedom to a group, which governs for the benefit of the individuals. The individuals choose their government and agree to abide by a set of rules, while the latter serves as an agent. The system starts with the interests of the individuals and ends with the individuals, leading to a democratic society. Nevertheless, democracy works best if the majority is well educated and not excessively selfish. Absolute freedom and total equality are mutually exclusive; in a democratic society a compromise between the two is necessary.

12.10 Foundations of Moral Code and the Human Nature

Following are some theories on the source and nature of moral code:

12.10.1 *Aristotelian view*

The purpose of life is to pursue happiness, which can be achieved only by rational thinking and following the middle course (golden mean) of action — neither excessive nor too little.

12.10.2 *Kantian view*

Morality is a "categorical imperative," meaning that it is a universal goodwill intrinsic to humans, an end in itself irrespective of the consequences.

We all have a social conscience — an inborn sense of what is right and wrong. Unable to find a simple explanation, Kant attributed the source of this intrinsic character to God. However, ever since the discovery of evolution, a straightforward explanation can now be derived from natural selection alone. The reason is simple: a species with an innate "conscience" that promotes group cohesion will have a better chance of survival than one without it. Kant died in 1804, fifty five years before the publication of Darwin's *Origin of Species*. Kant might have had second thoughts on the source of morality had he heard of the theory of evolution.[43]

12.10.3 *Judeo-Christian-Muslim view*

Unlike Kant who started from secular reasoning and arrived at God as a source of morality, the monotheistic religions assert at the outset that morality comes as a divine mandate, a decree from a personal God, as in the Ten Commandments. They adhere to the Golden Rule — do unto others what you would like others do unto you — and stress eternal happiness in the afterlife as a reward for a virtuous earthly life. Christians, in particular, attribute the negative aspects of human behavior to an "original sin" committed by man's forefather, and believe in the power and grace of God as the only way to absorb this sin and to achieve salvation.[44]

12.10.4 *Utilitarian view*

Advocated by John Stuart Mill, it stresses that the moral value of an act depends on its consequences in a society. From the standpoint of evolution, there is no intrinsic conflict between utilitarianism and the innate goodness of human nature, since our sense of goodness (conscience) is a product of our animal instinct, which in turn is shaped by the survival value of our species (utilitarian) selected over evolutionary time.

12.10.5 *Confucian view*

It believes people are basically kind and good, and they become bad only if they are exposed to bad external influence. It stresses the middle of the road to achieve personal happiness. Peace and harmony for all mankind is the ultimate outcome of a virtuous society. It teaches the equivalence of the Golden Rule — do not do unto others what you do not want others do unto you — without invoking a supernatural lawgiver.

12.10.6 *Buddhist view*

It stresses compassion not just for human beings, but also for all sentient living things. The reward for a virtuous life is to advance to a higher level of existence in the next cycle of transmigration. As is true with other Indian thoughts, the merits of good action are accumulated in the Karma system, taken as an impersonal cosmic principle.

What strikes me from these multiple views is not their differences but their similarities. There are glaring commonalities among peoples of varied racial origin, ethnicity, and culture, no matter which corner of the Earth they inhabit, and these are: respect for life, love and compassion, mutual help, consideration for others, avoidance of extremes of action, and the principle of reciprocity — all of which are requirements for the unity of a society. It should not come as a surprise that groups with these moral norms were favored by natural selection and survived the test of time. Successful groups were shaped by evolution to have these characters deeply ingrained in their genome and expressed as inborn conscience. It is therefore obvious that morality can be adequately explained by secular reasoning alone, without having to invoke the command from a supreme, supernatural being. Even the utilitarian view does not contradict the other theories, because a virtuous society is one that will deliver happiness to the people. Happiness, of course, is a catchy word. What is it really? In concordance with the theme of this book, I define happiness as the attainment of the goals of *self*: being able to exist and act, to freely express, to perpetuate, and to procreate without hindrance.

Morality stems from the relationship between individuals and the group, the balance between self-interest and group interest. A pro-social act is moral, and an antisocial act is immoral. Since a basic conflict exists between individual freedom and social cohesion, we all have antisocial as well as pro-social instincts. The adjustment and negotiation between *self* and mega-*self* is normal and necessary, and should not be taken with a sense of guilt. Finally, in a working society, most members are moral (pro-social) by definition, otherwise the society would have imploded and perished.

12.11 On Criminal Justice

What steps, then, should a society take against that small number of antisocial members? The answer is, if their action endangers a society, they should be segregated from that society for the protection of others. Education (including psychotherapy) and medication (if appropriate) should be tried. Punishment as an act of deterrence should be discouraged. Incarceration can be used as a means of segregation, although prison inmates should be made productive members of society by work, rather than becoming financial parasites. Capital punishment is the least desirable measure. Not only it is no more effective than other forms of punishment, it also leaves no chance for the offender to repent or to be exonerated. (Judging from the many cases of wrongful verdicts that were overturned by DNA testing, the merit of capital punishment is greatly in doubt.)

Although in principle criminal justice is straightforward, the implementation is complicated. I will not further elaborate but rather leave the details to the experts. Nonetheless, there is one point I like to clarify as it implicates the fundamental concept of *self*. This is about *human acts* as opposed to *acts of man*. The law defines a human act as one that is executed with the conscious intention of the agent (a person), whereas an act of man is one without this intention, such as by accident or by mistake. The question arises as to whether an action carried by a person with a "sick brain" is considered a human act. In fact, defense criminal lawyers tend to ask for excuse if their clients have a demonstrable brain illness, such as a tumor detected by an MRI. The controversy becomes more troublesome as diagnostic tests become more and more refined, and previously undetectable brain lesions become observable. The problem stems from the outdated Cartesian concept of mind-body dualism, which posits that mind is the "soul" that occupies the driver's seat in the brain, steering the brain (a machine or vehicle) to move and carry out actions. The conclusion derived from this assumption is that any malfunctioning of the "machine" is not the fault of the "driver." It can be argued, then, that a person with a brain tumor should not be held responsible should he commit a crime. The problem is, not all brain abnormalities are as clear-cut as a tumor. Many are subtle structural changes that may not be detectable today but will be tomorrow. Some brain illnesses are due to chemical changes, which are even harder to test, but they are abnormal nonetheless. To distinguish a normal from an abnormal brain could be a hair-splitting futile exercise. In my opinion, *self* is the agent of action, and *self* is a composite of mind and body, one inseparable from the other. A dysfunctional mind is also a dysfunctional brain, and *vice versa*. To acquit a criminal who has a demonstrable sick brain, and punish one whose brain illness is not easy to detect, does not sound like a fair judgment. My suggestion is to handle all antisocial persons as having a pathologic brain. They all have to be treated and they all have to be segregated from society until they are no longer a threat.

12.12 War Needs No Excuses

As long as there are *selves*, competition for survival is inevitable. War is a prototype of struggle among mega-*selves* — group selection at work (Figs. 12.2–12.4). In the animal world, competition for energy to sustain life comes in different forms: fighting for food; eating one another for nutrition; and territorial expansion for increasing food supply. The last is achieved by killing of competing groups of the same species, since animals of the same species consume similar food.

The Expanded Self: Society as Self   285

Fig. 12.2. Conflict within the same species in ants. Two groups of ants are entangled in a ferocious battle. [Courtesy Catherine Chalmers.]

Fig. 12.3. Ancient human warfare (Assyrian battle scene, 728 B.C.E.) [Permission British Museum.]

From archeological evidence to recorded history, almost all human societies have fought one another intensively and constantly, regionally and nationally. Starting from the hunter-gatherers, over 90% of human groups engaged in battle in one form or another.[45] Most human wars can be traced, directly or indirectly, to dispute over

Fig. 12.4. Modern human warfare (World War II). Battleship hit by a bomb. [US Natl. Arch.]

territory, including fertile farmlands, hunting grounds, and herding fields. War is also a means to capture slaves (as the Romans did), a source of manpower and energy. Other human factors, such as ideology, religion, and cultural differences, intensify and add layers of complexity to the conflict. Unfortunately, advancement in civilization does not mitigate the situation, but only magnifies the scale of war. Fueled by population explosion and dwindling resources, fighting among nations or human groups goes on. "A war that ends all wars" has never materialized.

Almost without exception, great conquerors of the past were ruthless killers. Westerners speak of Genghis Khan and Attila the Hun with terror, but Alexander the Great, much revered in Western culture, was no more merciful. He ravaged Persepolis in Persia to ruin. According to A. W. Benn, Alexander was "arrogant, drunken, cruel, vindictive, and grossly superstitious. He united the vices of a Highland chieftain to the frenzy of an Oriental despot."[46] Such was Alexander the man, despite two years of tutelage under Aristotle.

12.13 Intra-species Conflicts From Ants to Humans
12.13.1 *Ant warfare*

Ants are the most warlike of all animals. They conduct wars with colonies not only of alien species but also of their own. They far exceed humans in aggressiveness and cruelty. Their war policies include restless aggression, territorial conquest, and genocidal annihilation of neighboring colonies. Some ant species have a specialized soldier caste whose only role in the society is to fight and kill. They are larger than the worker ants and are equipped with strong muscles and sharp, clipper-like mandibles ready to chop off the enemy's head and legs, and to cut up the rest of its body into pieces. Some ants are programmed to be "suicide bombers"; their bodies are filled with toxic chemicals, ready to be discharged by an explosive muscle contraction when they are hard pressed during an attack or defense. Ant warfare is about territory and food, and is especially intense in times of food shortage. When worker ants contact an enemy, they run back to the nest, leaving a trail of pheromone to guide the recruits to the battleground. Their war tactics are extremely varied and include such ploys as stone dropping for attack and nest blocking for defense. Some species spray a chemical maze to disable and confuse the opponents in order to prey on their brood. Some species are expert sappers, digging elaborate tunnels from the home nest to the enemy nest before embarking on an all-out surprised attack. They also exercise rudimentary diplomacy by picketing the rival territory or bluffing their foes into submission without a fight. Some ants prefer to engage in an elaborate tournament, displaying their physical strength to intimidate and chase away the opposite party. The slave-maker ants capture their enemies as spoils of war and drive them into slavery.[47]

12.13.2 *Chimpanzee territorial expansion*

Perhaps the most vivid human-like warfare has been observed in chimpanzees at Ngogo, in Uganda's Kibale National Park.[48] These animals lived peacefully most of the time, but once every ten days, a band of males (up to twenty) would silently form a single file and cautiously patrol the border of their established territory. While doing this, the patrolling troop looked for males of the rival camp in the neighboring area. If the group was clearly out numbered, it would immediately disband and hurry back to its home turf. But if the soldiers encountered a lone male, they would kill it. The killing incidents were observed at least eighteen times over a period of ten years. Interestingly, most of the patrols occurred in an area to the northeast of their territory. After a decade-long "campaign," the Ngogo chimpanzees eventually annexed this new region, thereby expanding their territory by 22%, a definite selective advantage as they now had access to more food resources. The success of territorial expansion was an outcome of not only intergroup competition and aggression, but also internal cooperation and cohesion, being able to hold an unusually large community of 150 members (three times the usual size). What is biologically interesting is that, despite inter-group hostility, there was constant genetic blending between the rival communities (by grabbing each other's females), suggesting that in this instance group selection and not kin selection was at work.

12.13.3 *Rwandan massacre*

In Rwanda in the year 1994, the Hutu tribe committed genocide against Tutsi, killing approximately 800,000 in a hundred days. Both groups were Christians and they spoke the same language. The historical background is that, before Rwanda's independence, the Belgian colonists favored Tutsi as the superior tribe, using it to rule over the majority Hutu. Because of their superior position, the Tutsi were rich, whereas the Hutu were poor. On the other hand, the Hutu despised and hated the Tutsi and viewed them as foreign invaders who came to the land generations earlier. After independence in 1962, the power shifted from Tutsi to Hutu.

In April of 1994, extremist Hutu militia and elements of the Rwanda military decided to exterminate the Tutsi. The militia set up road blocks and slaughtered men, women, and children on the basis of their ethnicity. Radios incited ordinary Hutu to kill Tutsi, giving them directions as to where to find them and where they were hiding, and promising them the land of the Tutsi they killed. Many took up machetes and joined in the massacre. "Neighbors hacked neighbors to death in their homes, and colleagues hacked colleagues to death in their workplaces. Doctors killed their patients, and school teachers killed their pupils."[49] On 11 April alone, thousands of Tutsi met their deaths under the machetes of Hutu. By the end of June, when the bloodbath was halted, one out of eight people in Rwanda perished. Fearing retribution, Hutu militia fled to neighboring Zaire, taking with them 2 million Hutu refugees, many of whom participated in the slaughter.

What led to the ethnic feud over generations between Hutu and Tutsi, which culminated in the 1994 genocide? According to one analysis, the root of the trouble was that Rwanda is the most densely populated nation in Africa (eight million crowded into a small land), and the per capita tillable land has been strained to the limit. The fighting was really over who should have better control of land resources.

12.13.4 *Nanjing massacre*

One of the most horrific human intra-species conflict and atrocity happened in the year 1937 during the Japanese invasion of China. The incident was unmatched in history in the monstrosity of cruelty as well as in scale and duration. On 13 December of that year, the Japanese army marched into Nanjing (formerly known as Nanking), then the capital of China. In the ensuing six weeks, the Japanese soldiers, by order from their superior, systematically terrorized the city by engaging in intended mass torture and massacre of helpless civilians, and in collective raping of women, an incident later known as "The Rape of Nanking." The orgy of violence included large scale beheading of city dwellers in an assembly line manner. The captives were lined up in rows, and those in one row dug the mass grave for the next. Those still breathing were finished up with multiple bayonet stabbings, or were given to the dogs to be bitten and torn and eaten. Some were disemboweled or dismembered alive. Some were nailed to boards and run over with tanks. Some had their eyes gouged out and their noses and ears hacked off before being set on fire. One soldier, who participated in the mayhem but later repented, confessed sixty years after the event: "Few know that soldiers impaled babies on bayonets and tossed them still alive into pots of boiling water. They gang-raped women from the ages of twelve to eighty and then killed them when they could no longer satisfy their sexual demand. I beheaded people, starved them to death, burned them, and buried them alive, over two hundred in all. It is terrible that I turned into an animal and did these things." But how could a person turn into a beast? Another former soldier wrote, "Good sons, good daddies, good elder brothers at home were brought to the front to kill." Reluctant new conscripts were desensitized to cruelty and conditioned to kill by supervised practice on live victims with swords and bayonets, frequently in the form of contests, until atrocity became routine and banal, and human suffering, an amusement and recreation. By one account, about 300,000 Chinese civilians were slaughtered, a figure that matches the death toll from the atomic blasts at Hiroshima and Nagasaki combined. Among these, 20,000 were rape victims, who frequently were tortured to death or were killed following the sexual assault in order to eliminate evidence of the crime.[50,51]

What was the cause of these heinous acts? It appears that the tactic of torture and massacre was used deliberately to intimidate China to surrender. And behind all these acts was Japan's ambition of territorial expansion into the vast land of China, a greed for resources.

12.13.5 *Bataan Death March*

Albert Brown, the last survivor of the infamous Bataan Death March during World War II, died in 2011 at the age of 105. Brown, an American soldier, was sent to the Philippines in 1937 as an army captain. When the Japanese invaded the Philippines at the end of 1941, the outnumbered American and Filipino forces retreated into the mountainous jungles of Bataan, where they surrendered four months later following intense fighting. With little food or water and stricken with sickness, the emaciated soldiers were forced to trudge under the torrid topical sun to a prisoner-of-war camp sixty miles away. In the eye-witnessed account of Brown, those who could not continue the walk were killed on the spot. Japanese soldiers fractured their skulls with rifle butts and cut off their heads. Prisoners who tried to help fallen comrades were stabbed or bludgeoned. In the P.O.W. camp, survivors were listed in groups of ten. If one escaped out of the ten, they eliminated the rest of them. "So at night," said Brown, "just before roll call, you tried to find out if your ten were still there."[52] The Bataan incident occurred during the beginning of World War II in the Pacific, when Japan embarked on an all-out conquering of Southeast Asia, in an attempt to dominate the region.

Many Westerners find it particularly hard to reconcile the wartime cruelty of the Japanese military with the exquisite manners and politeness of peacetime Japanese people. Those who have had the experience of living in Japan post World War II were greatly impressed by their civility and social order.[53] It is puzzling that the same people who committed the Rape of Nanking and the Bataan Death March also exhibited the kind of altruism and empathy during the post-tsunami nuclear accident. The answer, in a nutshell, is *group selection*, in which intra-group compassion and inter-group brutality may show up as two sides of one coin. These diametrically disparate behaviors are intricately entwined. Diehard patriotism or war crime could just be a matter of viewpoint — from within or outside a group. A culture that overstresses internal unity may predispose itself to outward hostility and animosity.

12.13.6 *Other war atrocities and coalitionary killings*

The list goes on and on without end. Even the most glaring ones could fill up many more pages of this book. I shall only mention a few: The Holocaust (mass extermination in World War II; Nazis against Jews); former Yugoslavia (in 1995; Eastern Orthodox Serbs against Bosnian Muslims); Darfur in Sudan (since 2003, Arabs against Africans, both Muslims); Yelwa in Nigeria (2004; Christians against Muslims). Among these, the genocide of the Holocaust stands out loud and clear. As numerous books and public media have been dedicated to this part of history, I shall not devote any more space to it.

In short, no race or nation has a monopoly on atrocity. Cruelty, as well as kindness, is in all of us. It is all a matter of *self* against *self*, or, more precisely in these instances, mega-*self* against mega-*self*. We are born with war in our blood.

12.14 The Biological Roots of War and the Only Way to Eradicate it

We tend to categorize people into "them" and "us." The "in-groups" are always friendly, trustworthy, loyal, and kind, whereas "out-groups" are hostile, despicable, treacherous, and should even be eliminated.[54] We invent mythology, legends and creation stories to justify the superiority of our own group, and claim it to be the chosen people of the Creator. We go to war for a cause, to be sure, always a "good" cause, always a "righteous" cause, always one worth sacrificing life for. Every human being considers himself peace-loving, yet wars of all sizes never end.

As a rule, intense inter-group competition requires strong intra-group solidarity, since unity within a society enables it to outcompete other groups. Each group is an expanded *self* that seeks its own chance of survival, dominance, and perpetuation, often at the peril of others. As long as humans are partitioned into groups, war will never end. Short of an invasion from extraterrestrials, we earthlings are not going to stop killing one another to annihilation, thanks to the power (self-defeating, in this instance) of group selection. Peace on earth can only be achieved by eliminating all group boundaries, thereby coalescing all peoples into one single, ultimate mega-*self* that encompasses all of humanity.

In this balance hangs the fate of mankind. We are all denizens of the same planet, our only habitat in this vast cosmos. Either we go down the road of self-inflicted extinction, using increasingly lethal weapons of mass destruction, or we build a paradise on Earth by wisely and safely steering the course of biological evolution.

Notes and References

1. I like to distinguish mega-self from the so-called "super organism", first coined by William Morton Wheeler in 1928 and recently revived by Holldobler and Wilson. The difference is that super organism refers to insect societies, the organizational principle of which is 90% instinctive, based mainly on chemical communication (pheromones), whereas my concept of mega-self encompasses all levels of living things, including humans, and it is on the latter that this chapter emphasizes. Human societies are much more complex, and much more flexible, than those of insects, with instincts playing only a small part. See: Holldobler B, Wilson EO. (2009) *The Superorganism*. Norton, New York.
2. Riehl C. (2011) Living with strangers: Direct benefits favour non-kin cooperation in a communally nesting bird. *Proc R Soc Lond B* **278**: 1728–1735.
3. Nogueira T, Rankin DJ, Touchon M, *et al*. (2009) Horizontal gene transfer of the secretome drives the evolution of bacterial cooperation and virulence. *Current Biol* **19**: 1683–1691.
4. Cordero OX, Wildschutte H, Kirkup B, *et al*. (2012) Ecological populations of bacteria act as socially cohesive units of antibiotic production and resistance. *Science* **337**: 1228–1231.
5. Koschwanez JH, Foster KR, Murray AW. (2011) Sucrose utilization in budding yeast as a model for the origin of undifferentiated multicellularity. *PLoS Biol* **9(8)**: e1001122. doi:10.1371/journal.pbio.1001122.
6. Margulis L. (1981) *Symbiosis in Cell Evolution*. W. H. Freeman, San Francisco.
7. Hamilton WD. (1964) The genetical evolution of social behavior. I, II, *J Theor Bio*. **7**: 1–16.
8. In genetics, "ploidy" refers to the number of sets of chromosomes in the cell. A cell is diploid if it contains two sets of chromosomes; those with only

one are called haploids. Normally, animal cells are diploids. Hymenoptera are unusual in that they have diploid females and haploid males, a phenomenon called "haplodiploidy."

9. Wilson DS. (1975) A theory of group selection. *Proc Natl Acad Sci USA* **72:** 143–146; Wilson DS. (1997) Altruism and organism: Disentangling the themes of multilevel selection theory. *Am Naturalist* **150** (suppl.): S122–S134; Nowak MA, Tarnita CE, Wilson EO. (2010) The evolution of eusociality. *Nature* **466:** 1057–1062.

10. Wilson DS, Wilson EO. (2007) Rethinking the theoretical foundation of sociobiology. *Quart Rev Biol* **82:** 327–348.

11. In simple cases, Hamilton's model has explanatory power, but in real life, which is complex and multivariate, it loses its elegance. E.O. Wilson's colleague C. Tarnita presented the following hypothetical situation. Suppose your cousin is drowning and you want to risk your life to save him, not knowing that he is competing with your brother for the same girl. If your cousin survives to marry that girl, you may be depriving your brother of having progeny to carry on his genes, which are closer to yours than your cousin's are. (Quoted with permission from Lehrer J. (2012) Kin and kind. *The New Yorker*, March 5 2012, pp. 36–42.) What this story illustrates is that Hamilton's model works in isolation, but, like many other biological situations, there are numerous variables, many of which are not known to us. Aside from kinship, other factors also contribute to insect eusociality, among them a defensible nest (territory), genetic mutation, and selection by environmental forces. See Nowak MA, Tarnita CE, Wilson EO. (2010) The evolution of eusociality. *Nature* **466:** 1057–1062.

12. In an experiment set up to test the importance of touch to a youngster, infant monkeys were given a choice between a wire shaped in the form of a body and one covered with a piece of soft, velvet cloth as their surrogate mother. Whenever threatened, the monkey would cling to the surrogate covered with a piece of soft cloth. See: Harlow HF, Zimmermann RR. (1958) The development of affective responses in infant monkeys. *Proc Am Philos Soc* **102:** 501–509; Harlow HF. (1958) The nature of love. *Am Psychologist* **13:** 673–685.

13. Both oxytocin and vasopressin are peptides of nine amino acids, with slight differences in sequence. Both have peripheral and central actions. Those

secreted by the posterior pituitary gland go to the blood stream and act on peripheral organs. In this capacity, oxytocin is a hormone that enhances uterine contraction during delivery and promotes milk flow during lactation, while vasopressin is a hormone that causes vasoconstriction and enhances water re-absorption in the kidney, thus the other name "antidiuretic hormone." However, what is pertinent to this chapter is their central action — they are both produced inside the brain and act within the brain to promote pro-social behavior.

14. Donaldson ZR, Young LJ. (2008) Oxytocin, vasopressin, and the neuro genetics of sociality. *Science* **322:** 900–904; Insel TR (2010) The challenge of translation in social neuroscience: A review of oxytocin, vasopressin, and affiliative behavior. *Neuron* **65:** 768–779.

15. Wang Z, Yu G, Cascio C, *et al.* (1999) Dopamine D2 receptor-mediated regulation of partner preferences in female prairie voles (Microtus ochrogaster): A mechanism for pair bonding? *Behav Neurosci* **113:** 602–611; Aragona BJ, Liu Y, Yu YJ, *et al.* (2005) Nucleus accumbens dopamine differentially mediates the formation and maintenance of monogamous pair bonds. *Nature Neurosci* **9**: 133–139.

16. Lim MM, Young LJ. (2004) Vasopressin-dependent neural circuits underlying pair bond formation in the monogamous prairie vole. *Neurosci* **125:** 35–45; Lim MM, Wang Z, Olazábal DE, *et al.* (2004) Enhanced partner preference in a promiscuous species by manipulating the expression of a single gene. *Nature* **429:** 754–757.

17. Takayanagi Y, Yoshida M, Bielski IF, *et al.* (2005) Pervasive social deficits, but normal parturition, in oxytocin receptor-deficient mice. *Proc Natl Acad Sci USA* **102:** 16096–16101.

18. Lee H-J, Macbeth AH, Pagani JH, Young WS 3rd. (2009) Oxytocin: The great facilitator of life. *Proc Neurobiol* **88:** 127–151.

19. DeDreu CKW, Greer LL, Van Kleef GA, *et al.* (2011) Oxytocin promotes human ethnocentrism. *Proc Natl Acad Sci USA* **108:** 1262–1266.

20. Anstey ML, Rogers SM, Ott SR, *et al.* (2009) Serotonin mediates behavioral gregarization underlying swarm formation in desert locusts. *Science* **323:** 627–630.

21. The only exception is that early familiarity diminishes sexual attraction — nature's way to avoid incest.

22. Asch SE. (Nov 1955) Opinions and social pressure. *Scientific Am* **193**: 31–35.
23. Gould SJ, Lewontin RC. (1979) The spandrels of San Marco and the Panglossian paradigm: A critique of the adaptationist programme. *Proc R Soc Lond B* **205**: 581–598.
24. Adolphs R. (2002) Neural systems for recognizing emotion. *Current Opin Neurobiol* **12**: 169–177.
25. Singer T, Seymour B, O'Doherty J, *et al*. (2004) Empathy for pain involves the affective but not sensory components of pain. *Science* **303**: 1157–1162.
26. Kupferschmidt K. (2013) Concentrating on kindness. *Science* **341**: 1336–1339.
27. Rizzolatti G, Craighero L. (2004) The mirror-neuron system. *Ann Rev Neurosci* **27**: 169–192; Keysers C. (2010) Mirror neurons. *Current Biol* **19**: R971–R973.
28. For ethical and technical reasons, these human studies were limited to functional MRI and EEG. Thus the results are less clear-cut and precise compared to single cell electrical recordings conducted in monkeys.
29. Langford DJ, Crager SE, Shehzad Z, *et al*. (2006) Social modulation of pain as evidence for empathy in mice. *Science* **312**: 1967–1970.
30. Perez EC, Elie JE, Boucaud IC, *et al*. (2015) Physiological resonance between mates through calls as possible evidence of empathic processes in songbirds. *Hormones and Behav* **75**: 130–141.
31. Bartal IB-A, Decety J, Mason P. (2011) Empathy and pro-social behavior in rats. *Science* **334**: 1427–1430.
32. *The Guardian* (UK), March 15, 2011; *BBC News*, March 17, 2011; *Apple Daily*, Hong Kong, March 18, 2011.
33. Allen J, Weinrich M, Hoppitt W, Rendell L. (2013) Network-based diffusion analysis reveals cultural transmission of lobtail feeding in humpback whales. *Science* **340**: 485–488.
34. Beja-Pereira A, Luikart G, England PR, *et al*. (2003) Gene-culture coevolution between cattle milk protein genes and human lactase genes. *Nature Genet* **35**: 311–313; Ingram CJ, Mulcare CA, Itan Y, *et al*. (2009) Lactose digestion and the evolutionary genetics of lactase persistence. *Human Genet* **124**: 579–591.
35. Wrangham R. (2009) *Catching Fire: How Cooking made us Human*. Basic Books, New York.

36. Diamond J. (1997) *Guns, germs and steel*, Norton, New York.
37. Jablonka E, Lamb MJ. (2005) *Evolution in Four Dimensions*. MIT Press, Cambridge, MA.
38. One of the prominent examples of culture overriding genetics is the religious or ideological war waged among peoples of the same racial makeup.
39. Dawkins R. (1976) *The Selfish Gene*. Oxford Univ. Press, Oxford; 1989 edition, pp. 10–11.
40. My concept of "expanded self" differs from "meme" proposed by Dawkins. Whereas "meme" refers to human cultural artifacts such as fashions and ideas, my "expanded self" refers to groups that comprise individual members as in a human society. Dawkins excludes human assemblage as a type of "meme" because, had he done this, he would have contradicted his stand against group selection. See: Dawkins R. (1976) *The Selfish Gene*. Oxford Univ. Press, Oxford; 1989 edition.
41. Bekoff M, Pierce J. (2009) *Wild Justice*. Univ. of Chicago Press, Chicago.
42. In his writing "*Of Commonwealth*" in the book *Leviathan*, Thomas Hobbes (1588–1679) argued that human beings would live in harmony only if the laws of society are imposed by an awesomely powerful government — a "leviathan" or monster, without which discord is inevitable and society will collapse. John Locke's (1632–1704) idea of civil government is that people unite under a government in order to live comfortably and peacefully. Without a government, each person is free to enjoy his life and to punish those who harm him, but since others are equally free, there is no standard as to who should punish whom, and society will turn into chaos. Therefore, people are willing to sacrifice some of their freedom as long as the government can preserve their lives, liberty and possessions better than they can individually. See: "Of Civil Government" in The Second Treatise. See also: Berlin I. (1956) *Concepts and Categories*, Aristotelian Soc. Viking/Penguin. While Hobbes' view is consistent with an authoritarian monarchy, Locke's is that of a liberal monarchy. On the other hand, Jean-Jacques Rousseau (1712–1778), who believed in the innate goodness of human nature and the equality of all men, advocated liberal republicanism. His notion of social contract evolved into the ideal of a democratic society in which the supreme power is held by the majority of members who vote for representatives to govern them. See: Rousseau J-J. (1762) *The Social Contract*.

However, I should caution that even in a democratic society, the majority could, under certain circumstances, be persuaded by demagogues to vote for issues that turn out not to be in their best interest.
43. Kant initially rejected Thomas Aquinas' proofs of the existence of God based on reasoning (*Critique of Pure Reason*), but later affirmed God on the basis of moral necessity (*Critique of Practical Reason*).
44. The importance of the original sin is emphasized in the theology of Augustine.
45. Keeley LH. (1996) *War Before Civilization: The Myth of the Peaceful Savage*. Oxford Univ. Press, Oxford.
46. Benn AW. (1882) *The Greek Philosophers*. K. Paul, Trench, & Co., London, Vol. I, p. 285; as quoted by Russell B. (1945) in: *A History of Western Philosophy*. Simon & Schuster, New York, p. 160; see also: Durant W. (1939) *The Story of Civilization: Part II, The Life of Greece*. Simon & Schuster, New York.
47. Holldobler B, Wilson EO. (1994) *Journey to the Ants*. Harvard Univ. Press, Cambridge, MA.
48. Mitani JC, Watts DP, Amsler SJ. (2010) Lethal intergroup aggression leads to territorial expansion in wild chimpanzees. *Current Biol* **20**: R507–R508.
49. Gourevitch P. (1998) *We Wish to Inform You that Tomorrow We Will be Killed with Our Families: Stories from Rwanda*. Farrar, Straus, and Giroux, New York; as quoted by Chua A. (2003) *World on Fire*. Doubleday, New York.
50. Chang I. (1997) *The Rape of Nanking*. Basic Books, New York, pp. 58 & 59.
51. Shi Y, Yin J. (1997) *The Rape of Nanking*. 2nd ed. Innovative Publishing, Chicago.
52. Morrow D, Moore K. (2011) *The Forsaken Heroes of the Pacific War: One Man's True Story*. Tate Publishing, Oklahoma; Bedford Group, 2012.
53. Reid TR. (1999) *Confucius Lives Next Door*. Random House, New York.
54. Human nature assumes alien groups as dangerous and threatening, a phenomenon known as xenophobia. I am reminded of Stephen Hawking's advice that, should we encounter extraterrestrial intelligent beings, we had better avoid them before they would destroy us (ABC News, April 26, 2010). This echoes my suggestion that the default emotional response toward an unfamiliar stimulus is one of aversion (see *Chapter 9: Self and Emotion*).

Chapter 13 *Self* from Within: The Introspective *Self*

> *Present day physics will have to be replaced by new laws if organisms with consciousness are to be described.*
>
> — Eugene P. Wigner, 1963 Nobel Laureate in Physics

Overview: *Self can be approached from the outside and the inside. The observable self can be studied through biology, behavioral science, social science, and neuroscience. Cognitive neuroscience, the branch of neuroscience that deals with the behavioral outcome of the brain, is closest to the mind than other sciences, though it still relies on testimonies by the subjects about their inner selves but in fact not the inner feeling itself. The private, personal self is beyond the domain of natural science, and can only be reached by introspection. Introspective knowledge (such as the feeling of warmth rather than the rise in temperature) is less reliable and less reproducible, but nonetheless real and immediate. At the heart of introspection is the nature of mind and its relation with the body (or matter).*

Most philosophers today accept the reality of matter and the fact that the brain produces the mind, but the status of mind varies greatly. One school takes mind as an illusion. Another accepts the reality of mind but treats it as a bystander, having no effect on the brain. A third maintains that at some point mind and matter entwine and interact, a point where mind and matter "speak a common language." Currently there is no known natural law that can describe the interaction of mind and matter. The mind-matter conundrum remains an enigma.

13.1 "Look, How Happy the Fish Are!"

Master Zhuang and Master Hui strolled along the River Hao. Zhuang said, "Look, how happy the fish are!" Hui said, "You are not a fish.

How do you know the fish are happy?" Zhuang said, "You are not me. How do you know I do not know the fish are happy?"[1]

The above conversation, purportedly taking place two thousand and three hundred years ago, is to my knowledge the first recorded open discussion of the private nature of mind and how difficult it is to penetrate into other person's mind. What I will do in this chapter is no more than an extension of this protracted and endless contemplation on the most enigmatic part of our life. I will attempt to bring new insights into this age-old issue, taking advantage of the knowledge we have accumulated and the clearer perspective available of man's position in the universe. Perhaps I can dispel some of the enigmas, but perhaps I might make it even more mysterious.

The inner *self* is unfathomable and elusive, at least to a second or third person, yet it is also intimately real, tangible, and immediate to the person who calls himself "I." The key to the inner self is introspection. As William James once said, the state of consciousness by introspection "is the most fundamental of all the postulates of psychology."[2] Introspection is a process of reflecting on one's own mind. Introspective knowledge is private, but it is no less valid than other observable knowledge. In fact, introspection is a kind of observation, albeit directed inward and is limited to a single observer. It is not verifiable by other people and is not subject to statistical analysis. This shortcoming, hopefully, can be mitigated by the scrupulousness of the observed information, especially if made by a person who is keen and perceptive. It is the only window open if we are to look into our own mind. We have no other choice. As Descartes tried to say with his famous line, *"cogito, ergo sum"* (I think, therefore I am.), if we do not trust the existence of our mind, which gives us all we know, including the most tangible of matter, there is nothing else to be trusted.[3,4]

Charles Sherrington, after spending a lifetime as an illustrious experimental neurophysiologist, turned to the mind-world problem: "Between these two, perceiving mind and the perceived world, is there then nothing in common? Together they make up the sum total for us; they are all we have. We called them disparate and incommensurable. Are they then absolutely apart? Can they in no wise be linked together? They have this in common — we have already recognized it — they are both of them parts of one mind. They are thus therefore distinguished, but are not sundered. Nature in evolving us makes them two parts of the knowledge of one mind and that one mind our own. We are the tie between them. Perhaps we exist for that."[5] Bertrand Russell came up with a straightforward, witty, and somewhat cynical answer (attributing it to his grandmother), "What is mind? No matter. What is matter? Never mind."

Facing this daunting issue, there is a faction of scientists, neuroscientists included, who choose to disengage themselves from the topic of mind. The main reason is the "unscientific" nature of mind. These people take reality as something that can be experimentally confirmed. Some others are neutral but prefer to be mute on the topic. I can recall one instance in which a prominent neuroscientist publicly proclaimed the presence of mind but preferred to put it in cold storage since it is private and there was no way he could deal with it scientifically.[6] I also know of a number of neuroscientists who are attracted to their field by the mystery of the mind, hoping someday to be able to pry into its secrets, only to be disillusioned upon realizing that, as their own research gets deeper and deeper, their perspective of the mind gets hazier and hazier.

For those philosophically inclined, the mind-body dilemma continues to irk them. What, then, is the current status of the debate? Traditionally there are two extreme views: one holds that only mind is real; the other, only matter exists. With progress in natural science in the last three hundred years, we have learned to respect nature as an entity that stays whether or not we are here. After all, the world cannot be "wished" away. Therefore, with the exception of the very ill informed and those that are literal adherents of archaic religious dogmas, pure idealists are almost non-existent. Most respectable thinkers accept the reality of matter with varied views about the mind. The nagging question is: how much room is left for the mind, and what is the relation between mind and matter?

13.2 Three Views of the Mind

Today there are three main schools of thought regarding the concept of mind: (1) that mind does not exist — it is just an illusion; (2) that mind exists — it is a product of matter but it does not feedback on matter; (3) that mind exists — it is a product of matter and it feeds back on matter.

13.2.1 *Mind does not exist*

The first school denies the presence of mind — the private *self*. They take the physical changes in the brain as the only reality about the brain, and consider it unfounded for the separate existence of a private experience. For them, the study of the mind should be replaced by the study of the brain.[7] Daniel Dennett dismisses the subjective feelings (pain, for instance) as an illusion or misjudgment because they can be "explained away" by the physical processes observed inside the brain.[8]

The fallacy of this thinking is that explaining something does not necessarily negate its existence. There are many entities in the world that can be broken down (or "dissolved") into simpler, more fundamental components that are quite distinct from the original. Take a concrete example. A building can be mentally dismantled into bricks, cement, posts, beams, roof, and many other things, but this analysis does not annihilate the mansion standing before your eyes. Rather than being fictitious, the composite entity is the outcome of a phenomenon called emergence. *Emergence* refers to the fact that interactions of matter at one level, when reaching a certain degree of complexity (or a critical point), can lead to the appearance of new properties at a higher level (such as gaseous hydrogen and gaseous oxygen combining to form liquid water). Assuming the concept of "explaining away" were true, let us see what it would lead to. If the mental state does not exist because it is explained away by the physical state of the brain, then the brain processes can also be explained away by the molecular and ionic changes in the neurons, which in turn can be reduced into atoms, then to the elementary particles, and finally to the quarks and quanta, down to the hypothetical "strings" and "loops." (But wait, the reduction does not end here, for new theories are popping up to replace the matter/energy concept with even more fundamental "non-material relations.")[9] Thus, what the chain reaction of "explaining away" leads to is a vanishing point of unknown nature. In short, we either have to accept the phenomenon of emergence or to deny everything save a few abstract, nebulous hypotheses at the very bottom of the material world. But to accept the mind as an emergent is to recognize its reality. The problem with the eliminative materialists, therefore, lies not in the fact that the mind can be explained in physical terms, but in their assertion that after such explanation the introspective aspect of mind evaporates without a trace.[10]

It is true that scientific study of the brain will continue to progress, giving us insights into how the nervous system works and even a glimpse into how consciousness (as an observed phenomenon) can arise, not to mention the utility of such knowledge for the betterment of mankind in the control of neurologic and mental illnesses. It is also true that introspective examination of *self* does not reap such practical benefits, nor generate copious observational data to fill the books. Nonetheless, it is indisputable that the inner aspect of mind is there and will continue to be there, and for this reason alone, it deserves to be a part of philosophical discourse (if philosophy is taken to cover the entirety of human knowledge), though not necessarily a part of science. As Thomas Nagel once argued in his famous 1974 essay *What is it like to be a bat?*, the subjective *self* is at the core of our being, however "unscientific" it might be.[11]

Perhaps the best refutation for the "absence of mind" concept is not an argument, but a practical test — the "pinch test" as suggested by John Searle.[12] Go to those people who deny the presence of mind and pinch them hard in their arms unprovoked and without warning. If they get mad and chase you down the stairs, there is no doubt that they have an inner experience of pain (the mind) that can affect matter (the body) and set it into action.[13]

For those who still insist that our inner experience is illusory, let me make one more point. For an agent to have an illusion, there has to be an experience, the content of which is distorted. Therefore, denying the presence of mind yet accepting the experience of an illusion is self-contradictory.[14]

13.2.2 *Mind is a bystander*

The second school accepts the reality of mind as a subjective, first person experience, but takes the causal relationship between body and mind as a one-way street, that is, the body causes mind but mind does not cause the body. In this view, the mind is a "supervenience" of the brain. What this means is that mind is a bystander of brain activities and the brain's interaction with the world. It is like a sports spectator who participates in the tension, anticipation, excitement, ecstasy and despondency of a ball game, without himself ever causing the ball to roll. One subset of this school is the so-called "property dualism," which states that mind and matter both exist but only matter (the brain) is of real substance. Mind is just a "property" of matter, an epiphenomenon. The presence of mind is therefore fortuitous — we are just lucky to have it and, with it, be able to enjoy all the good things (or to suffer the pains) of life without having to do the work (the body does it).[15]

The main objection to this idea is that it does not make evolutionary sense. Mind is not only a biological phenomenon (a product of the life process); it also has adaptive value. Having a mind is advantageous in a competitive world, as the ability to foresee future events before they actually happen enables an animal to better cope with the challenges in life. In the process of evolution, a trait that is profitable to a species is "zealously" retained and one that diminishes its adaptability is mercilessly eliminated. Those in the middle ground (neither good nor bad) go through a phenomenon called "genetic drift," in which a neutral trait may appear or disappear randomly across evolutionary time. It certainly is not bad for animals to have minds. But if it were not essential for survival, it would not have staying power over eons of biological evolution. And for mind to be useful for survival, it has to be able to act on the body to influence the environment, making the mind-body relation a two-way traffic.[16,17]

13.2.3 *Mind interacts with body*

The above discussion automatically leads us to the third school, the one that looks appealing to me, which is very much like the second except that a bidirectional action between body and mind is considered possible. Although all the brain activities can be described in causal physical terms, and although these physical activities are closed, it remains possible for the mind (a product of the brain) to "somehow influence" the ongoing brain activities.[18] Now, an insurmountable problem arises: there is no known natural law that can describe the effect of mind over matter. How can mind, which has no mass/energy value and does not occupy space,[19] "jump-start" the molecular and cellular events in the brain? I have no answer. I can only offer three wild conjectures: (1) We currently have no conversion formula from matter to mind; perhaps once we have it, the reverse process could be better understood or described. (2) The absence of a law to connect two natural events is no reason to negate the presence of either. The mind-body "law" could be quite different from any other natural law we have ever come to know. (3) It might be possible that mind *emerges* at a certain stage of physical activities of the brain, and that at this critical point mind and matter become one and the same (a novel "state of matter," so to speak), so that mind activity and brain activity are inseparable.[20]

Perhaps we might have an answer in the future, but perhaps not. There are many occurrences in the world that we cannot yet satisfactorily explain; the nature of mind is just one of them, the other being the nature of matter. At the very bottom, gravity does not mingle with quantum mechanics, and this has all the physicists scratching their heads. Christof Koch, physicist turned cognitive neuroscientist, after years of tackling the problem of mind, transpired an air of humbleness:

> "I would like to end with a plea for humility. Humility because even though we are living in the age of science, we know so little. The

cosmos is a strange place. Take the decade-old discovery that only four percent of the mass-energy of the universe is the sort of material out of which stars, planets, trees, you, and me are fashioned. One quarter is cold dark matter while the rest is something bizarre called dark energy. Cosmologists have no idea what exactly this is nor what laws it obeys. It is exceedingly strange staff and cannot be seen. Is there some ephemeral connection between this spooky stuff and consciousness, as suggested by the novelist Philip Pullman in his trilogy "His Dark Materials"? Very likely not; but who is to say for certain. Our knowledge is only a fire lighting up the vast darkness around us, flickering in the wind. So let us be humble and be open to alternative, rational explanations."[21]

13.3 Do Other Beings Have Minds?

In taking up this topic, I shall limit myself to the simplest aspect of mind — consciousness. And in dealing with consciousness, I shall limit to its simplest function, that of awareness of the environment. For the sake of argument, I shall use the word "being" in this section to include not only living things but also the information-handling machines made of inanimate materials — computers and robots.

There are two extreme views on this subject: the solipsist's view that posits that nothing but myself has a mind; the panpsychist's view that believes that everything in the world has mind, including the inanimate matter. My own view is that the assignment of mind is a matter of *probability*, from the certainty of my own (probability of 1.00) down the evolutionary ladder to a probability approaching zero (though I cannot tell you where the zero point lies). I agree with Darwin's view that whether animals have minds is a matter of degree and not of kind.

That I have a mind is, to me, a given, but whether other people also have minds is a different story. I never doubt it, but I cannot be one hundred percent sure. I learned very early in life that other human beings, who look like me, talk like me, walk like me, and react with the same facial expressions to things that I like or dislike, also have minds.

But here, mind starts to lose its immediacy. For example, when my wife has an attack of migraine headache, she may not show it in her facial expression, and frequently I have to rely on verbal communication to know it. The same situation applies to people not in my household but whom I meet every day, such as my colleagues. When it comes to people that I have never met in person but have seen their images on television and heard their voices on radio, such as the President of the United States, I also have no doubt of their minds, but here the immediacy recedes further back. I also assign consciousness, with a fair degree of confidence, to historical figures — Plato, Beethoven and Kandinsky — who are not my contemporaries, but who through their writing and creative work I have a mental rapport with. Next come my household pets. Most pet owners treat their own dogs and cats like family members, imbuing them with a mind, but (ironically) the same persons might treat other people's pets as if they have no consciousness. I once owned a dog and we got along well by tacitly respecting each other's mind — temperament, idiosyncrasies, habits, likes, and dislikes. My dog understood twenty some spoken words but not sentences. She had some rudimentary reasoning. I assumed she had a mind, but it stayed at a level of a two-year-old child. Once I met a bonobo (a species of chimpanzee with a high level of intelligence) who was proficient in using computer icons in communicating with humans, and who also learned how to make and use tools.[22] I believe its mental capacity is higher than that of the dog but still lower than that of a grown-up man. How about the invertebrates — the bees, ants, and scorpions? They have very limited reasoning, depending highly on impulses driven by instincts and pheromones, but I am inclined to take them as conscious, as they respond quickly to stimulus from the environment. How about the unicellular eukaryotes, those that can only be seen under the microscope — the amoebas, the parameciums, and the stentors? Under observation, amoebas seem to "know" what they are doing. Some biologists like H.S. Jennings (see *Chapter 4*) think they are conscious. I simply do not know; only an amoeba knows, if only it has a mind to know. When we go down to the bacteria, the controversy increases, but the possibility of a trace of awareness lingering on cannot be ruled out. No doubt consciousness fades gradually down the evolutionary ladder, but at what point it vanishes is anybody's guess. On the other hand, it might be possible, though not provable, that atoms and molecules in the inanimate world might harbor a dormant form of mind, waiting to germinate into consciousness when conditions are right (when they become part of a living system). In short, I like to assign consciousness in its absolute term only to myself, and impart *relative* consciousness to all other beings.

Now we come to the thorny question of whether computers and robots have minds. I shall confine my discussion mainly to computers, as robots are just computers with moving parts. There is a school that maintains that anything that shows "intelligence" has consciousness — the so-called "strong artificial intelligence." (By intelligence, they mean a problem-solving logic, or an algorithm in computer parlance.) According to this school, not only do computers and robots have minds, but thermostats also do.[23] They consider a thermostat as a small robot, a regulator with a feedback mechanism. Their reasoning goes like this: computers simulate certain aspects of human thinking, and since our thinking organ (the brain) is capable of awareness, computers should also be conscious.[24] My main objections to this school are: (1) If any programmable machine has consciousness, why not assign consciousness to any algorithm, including one that is written on a piece of paper? (2) The brain does not process information *perfectly* logically. It errs, and it is tainted by emotion, passion, desire and will. The brain has an appetitive as well as rational function. Awareness of something without the accompanying liking and disliking does not make biological sense. (See *Chapters 8 and 9* on the interdependence of consciousness and emotion.) (3) The "intelligence" of a computer or robot is simply the extension of the intelligence of the hardware maker and the programmer. There is no need to impart an independent existence of consciousness to the computer.[25] (4) A computer, even a self-reproducing robot, does not have the self-generated propensity to seek its own perpetuation;

it will stop somewhere when it runs out of components, raw materials, or energy source (see Fig. 2.3 in *Chapter 2: An Astronaut's Dilemma*). What a computer lacks, but a blade of grass has, is *self*. (5) The brain is such a complex, organized piece of matter that no simplified version may duplicate its function, let alone the generation of mind. I suspect that to come up with a mind, nothing short of making another brain can do the job.

Lastly, let me bring out the caveat that consciousness is a private business that cannot be experienced by a second party. Since I am not a computer, I cannot tell you with certainty whether I (in this instance a computer) am conscious. Thus I cannot assert that the probability of consciousness for a computer is absolutely zero. I can only say that it is very, very low indeed — almost zero or approaching zero.

13.4 Can Mind Stand Alone?

Without matter, mind cannot stand alone. The brain is such a strange organ that it always projects or refers outward. For example, when it executes an action, something (e.g., muscle) other than the brain moves. When it feels, the sensation falls in other parts of the body and not in the brain. Thus the brain itself is devoid of motion and sensation. The cognitive function of mind requires an external world to recognize, while its affective faculty requires a body for emotion to be expressed. Without input from the sensory nerves a person can never perceive or imagine a flower. Even the memory of abstract numbers needs concrete matter for the numbers to anchor to. Mind reflects physical reality, without which mind has no content. Just as a mirror that reflects nothing ceases to be a mirror, pure mind without matter ceases to be a "mind." In short, mind alone is a non-entity.

13.5 Can Mind be Explained in Physical Terms?

There are three approaches to a physical explanation of the mind: (1) correlative explanation; (2) causal explanation; (3) ontological explanation. The first two approaches are easy to understand, but the third is not. Correlative explanation is easy to do and has been done many times over. Very roughly, we can correlate the brain with mind, for if we remove the brain, the mind (at least its outward behavior) is gone.[26] We can also correlate the electrical phenomenon of the brain with mind, for we can identify the degree of alertness with the frequency and amplitude of the brain waves. Furthermore, we can correlate mood changes with brain chemistry, because if we take a certain drug that is known to affect synaptic transmission, we become happier or sadder. The examples are endless and are accumulating continuously. Causal or mechanistic explanation goes a step further. As techniques become more refined and analyses more sophisticated, we can get closer and closer to the abode of mind. For example, analysis of the brain used to be limited to the human organ taken after death, or to a non-human animal, but now electrical activities and chemical changes can be studied in a live human brain. Instead of using a single electrode, multichannel recordings on live, conscious persons are being done. Non-invasive imaging techniques and genetic manipulations can pinpoint functions in small brain parts. Indeed, progress in physical study of the mind will continue with no end in sight. Nevertheless, while we rejoice in this type of explanation, the question arises as to when we will ever reach the "crossover" point (a threshold) where the physical turns into mental, identified introspectively as our inner feelings, thoughts, desire, and of course our sense of *self*. But think of it carefully, is there really such a point? The matter-energy world seems to be a closure. Is there a point at which matter "jumps over" to the mental? The third type of explanation, ontological, is a philosophical issue outside the domain of science, and by far the most difficult to tackle.[27] What is the nature of mind? How do we interpret the nature of mind in terms of the nature of matter? The nature of mind is irreducible, as is the nature of matter. In fact, we do not even know the nature of matter; how can we use it to understand mind? The mystery of mind plus the mystery of matter is the mystery of everything, or the mystery of our being. As matter and mind are distinct yet intertwined, could the two mysteries be just two sides of one coin?

13.6 Afterthought: A great paradox

We live in an intuitive world of harmony and coherence. It is only when we start to analyze that the dichotomy of mind and body emerges, and this plunges us into an entangled mesh of physical and mental realities. The material world is real, but its nature is forever hidden behind a "veil." The mental world is also real, but because we dwell in it, and it is in us, it is always ineffable and indescribable. Mind and matter are different but inseparable. It may be more correct to say that there is only one real world, the intuitive world that is both mental and physical (or neither totally mental nor totally physical). Facing this enormous paradox, I can only admit that the mind-body conundrum, which is the very essence of our being, remains an enigma.[28]

Notes and References

1. Excerpted from Chapter 17 of the Chinese philosophical classic *Zhuangzi (Chuang–Tzu)* 《庄子》. Original text: 庄子与惠子游于濠梁之上。庄子曰: 儵鱼出游从容, 是鱼之乐也。惠子曰: 子非鱼, 安知鱼之乐？庄子曰: 子非我, 安知我不知鱼之乐？
2. James W. (1890) *The Principles of Psychology.* Henry Holt; republished by Dover, New York, 1950, Vol. 1, p. 185.
3. Descartes R. *The Philosophical Works of Descartes.* English trans. by Haldane ES, Ross GRT. (1967) Vol. 1, page 101, Cambridge Univ. Press, New York.
4. Note that Descartes used the term "cogitation" to encompass all mental functions, including thinking and emotion. As Bertrand Russell said, "Thinking is used by Descartes in a very wide sense. A thing that thinks, he says, is one that doubts, understands, conceives, affirms, denies, wills, imagines and feels — for feeling, as it occurs in dreams, is a form of thinking." See: Russell B. (1945) *A History of Western Philosophy.* Simon & Schuster, New York. Descartes' famous statement (cogito) represents an epistemological approach to affirm the existence of the agent who does the mental activities. The approach provides a solid foundation for the assertion of *self* and mind. It is only when he imbued the mind with immortality (soul) and

independent existence that controversy arises. It is therefore important not to confuse the two issues — the "presence of mind" and the "immortality of mind."

5. Sherrington C. (1940) *Man on His Nature*. Cambridge Univ. Press, Cambridge, UK; revised 1951, Mentor Books, New York, Chap. 11.
6. Kety SS. (1960) A biologist examines the mind and behavior. *Science* **132**: 1861–1870.
7. Churchland PS. (1986) *Neurophilosophy*. MIT Press, Cambridge, MA, page ix; Churchland PM. (2007) *Neurophilosophy At Work*. Cambridge Univ. Press, New York.
8. Dennett D. (1991) *Consciousness Explained*. Little, Brown & Co., Boston, pp. 454–455.
9. Kuhlmann M. (2010) *The Ultimate Constituents of the Material World: In Search of an Ontology for Fundamental Physics*. Ontos Verlag, Frankfurt; Kuhlmann M. (August 2013) What is real? *Scientific Am* **309**: 40–47.
10. In more recent writings, Patricia Churchland seems to be backing down from the position of rejecting mind as reality. Nevertheless, she still insists that the philosophy of mind should be replaced by the science of the brain. See: Churchland PS. (2002) *Brain-wise: Studies in Neurophilosophy*. MIT Press, Cambridge, MA, pp. 2 & 401.While I agree on the promise of brain science, I do not believe that it can totally wipe out the philosophy of mind.
11. Nagel T. (1974) What is it like to be a bat? *The Philosophical Rev* **83**: 435–450.
12. Searle JR. (1992) *The Rediscovery of the Mind*. MIT Press, Cambridge, MA; Searle JR. (1997) *The Mystery of Consciousness*. New York Rev. of Books, New York.
13. Pain is defined as an intensely disagreeable sensation.
14. The manifestations of consciousness are so obvious and prevalent in higher animals that it would be almost impossible to imagine what a "mindless" person would be like. The romantics may picture a "zombie," but nobody knows what a zombie really is, since it is a non-existent creature. The medically minded can compare it to a person suffering from sleepwalking, but no one can engage in sleepwalking for more than a few minutes; and I do not believe how such a person could spend a lifetime undertaking such

complex activities as attending college, flying a Boeing 747, partaking in a chess tournament, performing the Rachmaninoff Piano Concerto No. 2, and perhaps even stealthily designing a nuclear bomb.

15. Chalmers DJ. (1996) *The Conscious Mind.* Oxford Univ. Press, Oxford; Edelman GM, Tononi J. (2000) *A Universe of Consciousness.* Basic Books, New York.
16. I owe this evolutionary insight to a discussion with Professor Francesco Orilia. Also, the argument from evolution was presented earlier by Karl Popper. See: Popper KR, Eccles JC. (1981) *The Self and Its Brain*, Springer, Berlin, p. 72.
17. It can be contended that, in evolutionary history, mind appeared as a by-product of other functions essential for life, and that mind subsequently persisted because it turned out to be useful. This argument is akin to the biological "spandrel" theory of Stephen Gould, which I refuted in *Chapter 12: The Expanded Self*. My reason is that if a new trait turns out to be of adaptive value, it will be selected and retained whether or not it arises by accident. In fact, most mutations arise by accident without biological "foresight". It is the adaptive outcome that decides the fate of a trait. For Gould's argument, see: Gould SJ, Lewontin RC. (1979) The spandrels of San Marco and the Panglossian paradigm: A critique of the adaptationist programme. *Proc R Soc Lond B* **205**: 581–598.
18. Perhaps the most glaring example of mind affecting the body is when a person commits suicide. In this case mind destroys the body. I owe this insight to David Noerper.
19. Although mind seems to have a dimension of time.
20. The idea of a new "state of matter" may sound ridiculous, but it should not be dismissed as a total nonsense. An example is the state of Bose-Einstein condensate: When matter is cooled to near absolute zero degree, quantum effects prevail on a macroscopic scale. This example, however, is used merely to demonstrate that we should be open to unusual possibilities. It is not meant to endorse a quantum theory of mind.
21. Koch C. (2009) Free will, physics, biology, and the brain. In Murphy N, Ellis G, O'Connor T. eds. *Downward Causation and the Neurobiology of Free Will.* Springer-Verlag, Berlin, pp. 31–52. (Quoted with permission from Springer Berlin Heidelberg.)

22. Savage-Rumbaugh S, Lewin R. (1994) *Kanzi: The Ape at the Rink of the Human Mind.* Doubleday, New York; Savage-Rumbaugh S, Shanker SG, Taylor TJ. (1998) *Apes, Language, and the Human Mind.* Oxford Univ. Press, New York.
23. Chalmers DJ. (1996) *The Conscious Mind.* Oxford Univ. Press, Oxford.
24. The term "strong artificial intelligence" or "strong AI" was coined by John Searle to denote the idea that computer science can create real minds, as opposed to the notion of "weak artificial intelligence" or "weak AI," which maintains that computer science can at best simulate certain functions of minds but never becoming a real mind, such as what a human being experiences.
25. A similar situation occurs when a person watches a puppet show. One may be so absorbed as to believe that the puppet is alive with intentions and feelings, forgetting the fact that these human-like features are no more than the skillful manipulations of a puppeteer.
26. Hippocrates, over two thousand years ago, is said to have correlated the brain with mental functions such as thought, emotion, perception and choice.
27. "Ontology" is a philosophical term, which in common language refers to the nature of things. For example, my subjective feeling of hot or cold is not the same as the observed rise or fall of the mercury column in a thermometer. Though correlated, they are different kinds of reality.
28. My position regarding mind-body relationship is closest to that of "neutral monism," as defended by Bertrand Russell and others.

Chapter 14: *Self*, Realities, and the Transcendents

The attainment of a cosmic self is the highest goal of living.

Overview: *Humans have amassed a large amount of knowledge about themselves and the world, but there are limitations to human understanding. There are facts that we do not yet know or cannot yet explain. And still there are realities that, because of their fundamental nature, can never be clearly understood. Mind is one of these. Mind is the axiomatic standpoint from which we know everything else. It is simply there, beyond proof or disproof.*

Our reflective capacity leads to the realization of the finitude of life. Contrasted with the enormity of the universe, our ephemeral existence appears worthless. Religion is the question raised when our transitory self confronts the immensity of the universe. It is an endless search without a clear-cut answer. However, in the pursuit of this question, we are linking our minuscule self with the immense, and our transitory self with the constant. The merging of man and nature is the highest form of being. In this manner, self, which initially "crystallizes" out of the world, in the end yearns to return to it.

14.1 *Self* and the Realities

When *viewed from the inside*, mind is the indubitable, *supreme reality*. From this axiomatic foothold we process the *derived reality* of the outside world. However, it is only partially true to say that the mind reflects the world. It would be more correct to say that mind reflects part of the world. On the perceptual plane, each animal species lives in a world of its own, thanks to the presence or absence of certain sense organs. A dog, for instance, lives in a world of odor mostly oblivious to its master. A bat constructs the world with its ultrasonic radar machine. A shark maps changes in electrical potential in its surroundings. Humans, by contrast, rely very much on detecting, by sight, a limited range of the electromagnetic wave spectrum. Even within humankind, every person's world is somewhat different from those of others, being influenced by differences in attention, interest, emotion, experience and culture. (Note that people witnessing the same incident frequently provide conflicting reports.) Our nervous system sets a limiting condition for the reconstruction of the world out of the raw materials available in the environment. The raw materials in the physical world are presumably real and constant, but the derived physical world is unique to each person.

What is the reality of a tree falling? Does it really fall? If I actually see it, and hear it, falling, then my mental reality has a *primary construct* (out of raw sensory data) of the tree falling; i.e., I perceive its fall. If I do not witness it but hear someone reporting it, then I will have a *secondary construct* of a tree falling, a mental reality assembled by reshuffling pre-existing sensory data stored in my memory. Both instances are mental activities. Then, from my own mental reality, I project outwardly a physical reality in which the tree falls. The physical reality becomes a fact if the experience is shared with many people, or if the event can be verified again and again, especially not only by sight but also by touch. By now the objectivity of the tree falling is established. Once complete, such as when the event is reported in the newspaper, it finally becomes independent of my mental activity. But note that whatever the means of arrival at objectivity, the starting point is always the mental reality.

In everyday life, the hallmarks of reality are endurance and consistency. A phenomenon that is fleeting and cannot be observed a second time is less likely to be real than one that has staying power. Likewise, two observations that are contradictory cannot both be real. Dreams differ from time to time, and from one to another. Even in the same dream, events are usually inconsistent. Thus dreams are not real despite the fact that they are also a part of our stream of consciousness. In contrast, events observed in our waking hours are invariably taken as real, for they either stay the same day after day or can be logically connected from one day to the next.

The sense of one's *self* does not fluctuate widely from day to day. There is a famous parable in the book *Zhuangzi* (*Chuang-Tzu*). The story goes like this. One night Zhuangzi dreamt of being a butterfly, flitting and fluttering happily. Upon waking up in the morning, he was disappointed to be a man. He then questioned his real identity: was he a man dreaming he was butterfly, or a butterfly dreaming it was a man? We can judge for him that he finally realized he was a man, because being a man occurred every day while being a butterfly happened only once. But what if his butterfly dream came every night? Then he would have a hard time deciding.[1]

Scientific knowledge is an expansion of common sense. Scientists build instruments to extend the range of observation and extract information by correlating one variable with another while holding all others unchanged (the experimental method). In this manner scientists do a better job than ordinary people in using *derived physical reality* to fathom the hidden *primary physical reality*. Scientists build models to fit the experimental data. These models are based on an overall framework called "paradigm," which is to be changed (paradigm shift) when it is no longer internally consistent, as what happens during a scientific revolution (such as the Copernican revolution of planetary movements and the Einsteinian revolution of space–time and gravitational concepts).[2]

Ever since the optimistic vision projected by Francis Bacon in the seventeenth century, scientific knowledge has been expanding at an exponential rate, giving us a sense that science could one day solve all the problems facing humankind and answer all the questions man will ever ask. While this may be true with the utilitarian power of science, it may not be so with regard to the fundamental understanding of nature. Our model of physical reality is never final; it is subject to change and revision as new information arrives. Thus, there is always an epistemic

318 Self and the Phenomenon of Life

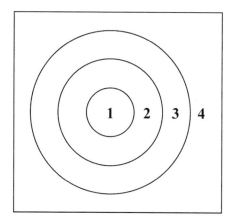

Fig. 14.1. The sphere of knowledge. Zone 1 covers facts that are known and explainable (interconnected). Zone 2 covers known facts that are not explainable (no connection to zone 1). Zone 3 covers facts that are currently unknown (but potentially knowable). Zone 4 covers realities that, because of their fundamental nature, are unknowable and unexplainable. (The circles are drawn to show only the relationship; no meaning is given to the relative size of the areas.)

limitation, or rather, uncertainty, to scientific inquiry. Perhaps it would be prudent to assume that the curve of scientific progress is asymptotic, ever increasing but never reaching an endpoint.

Figure 14.1 represents the totality of reality in terms of human knowledge. Zone 1 encompasses knowledge of events we see in everyday life. They are related by common sense and integrated with scientific theories and equations. Zone 2 contains events that are known and observable but not yet mathematically incorporated into Zone 1. They are the currently unexplained entities such as the "dark matter," the "dark energy," and the relation between quantum mechanics and gravity.[3] Zone 3 contains knowledge unknown to us but potentially will be known in the future. Since I cannot tell what it is, I can only provide a hint. Stephen Hawking, the astrophysicist, when asked what would be the greatest discovery in the next hundred years, predicted that it would be something that we least expect to discover. There is also the famous quotation, attributed to Isaac Newton, known to every high school student: "I seem to have been only like a boy playing on the seashore, and diverting myself in now and then finding a smoother pebble or a prettier shell than ordinary, whilst the great ocean of truth lay all undiscovered before me." Zone 4 is reserved for a special type of reality that, because of its fundamental nature, is forever inaccessible to scientific inquiry. Such reality provides the axiomatic starting point for science but is itself outside the reach of science. Questions such as the nature of the subjective side of mind and "how does anything exist at all?" belong in this category. Its presence is simply *self*-evident, but is beyond proof or disproof.

Scientists deal with knowledge in the first two zones, and attempt to extend into the third zone. Science aims at demystifying natural phenomena by offering explanations through principles derived from simple physical events (natural laws). Since scientists engage only in realities that are observable and testable, and regard all others as *technically* non-existent, within this confine, scientists can safely say or hope that there is no problem that cannot ultimately be solved. It should be recognized, however, that this attitude is only *operationally* correct; to extrapolate this to cover the entire realm of reality is without basis.[4]

Whereas scientists dwell most comfortably in the first two zones, philosophers, on the other hand, cover all and take the fourth category seriously as a subject of discourse. As scientists are concerned only with *testable* hypotheses, philosophers *raise questions* regardless of whether there will be an answer. Here lies a fundamental difference: science by definition deals only with positive knowledge; philosophy deals with both positive and negative — the knowable and the unknowable. For this reason, science, powerful as it is, cannot totally replace, or engulf, philosophy. The emphasis on the unknowable is particularly evident in Eastern philosophical thought. As Youlan Feng, the eminent Chinese philosopher of the twentieth century, pointed out, Chinese traditional philosophy contrasts with Western analytic thinking in being intuitive and holistic, and paying much respect to the unknown. Feng suggested that a balance of the positive and negative sides of knowledge should be the correct attitude of philosophy.[5]

A related issue is the difference between "science of mind" and "philosophy of mind." Both have mind as their subject matter, but the former, as exemplified by cognitive neuroscience, deals with mind as reported by an experimental subject (an externalization of inner feeling but in fact not the feeling itself), whereas the latter includes, in addition to the observable, the internal feelings and experiences which are private, ineffable, and cannot be shared by a second person. It is tempting for cognitive neuroscientists, and some science-leaning philosophers as well, to deny the reality of the inner mind and to take its externalized surrogate as all there is, and thus mistakenly equating the "science of mind" with the "philosophy of mind."[6]

There is also the mystery of the fundamental constants of nature. These are observations (measured and expressed in numbers) that are empirically obtained but defy explanation by simpler laws. They include Planck's constant, the speed of light, proton mass, electron mass, electron charge, and gravitation… all twenty-three of them, commonly listed on the inside cover of a physics textbook. These constants define the universe we live in and are critical for the existence of life. Any minor deviations from them would have led to a drastically different world and we would not be here to observe it.[7] It might be possible that, in the future, some of the constants could be combined and reduced to a simpler set of numbers, but not all will go away and a few will remain unexplained.[8,9] For the time being, I would place the fundamental constants of nature in Zone 4.

Aside from mind and matter, Roger Penrose[10] proposes a third reality — the mathematical reality. Mathematics is a product of the human mind, yet it does not exist exclusively in one person's mind. It is shared by many and can be verified independently over long intervals of time and across generations. If a person does not discover a certain mathematical principle, other people will, sooner or later. It has a certain objective, independent existence, despite being the product of the mind. It can be applied to the physical world, yet it is not a constituent of matter. It is a strange entity — part mind, part world, but neither one nor the other. The same can be said of natural laws, which are expressed in mathematical terms, and of logic, the foundation of mathematics. The formal nature of mathematics has an idealistic touch of Platonism.

Perhaps the correct philosophical stand is not monistic or dualistic, but a tripartite amalgam of mind, matter and form. John Barrow compared the mathematical laws of nature to the "software" running on the "hardware" of matter and energy.[11]

May I add that mathematics in a broader sense is not the monopoly of adult *Homo sapiens*. Human infants and other vertebrates (chimpanzees, birds, and rats, for example) also possess a mental "accumulator," a rudimental sense of magnitude that distinguishes "more" from "less," or "greater" from "smaller." From the evolutionary point of view, the "number sense" is a biologically adaptive function created by the brain to deal with the outside world. It is a bridge that goes between mind and matter.[12]

14.2 Can Science be Totally Objective?

Let me deal with the most objective of all sciences — the physical science. Physics depends on two basic assumptions: that the physical reality exists, and that the subjective idiosyncrasy of the observer can be suppressed. The objectivity of scientific data derives from the fact that they can be verified from one observer to another and at different times. However, it should be pointed out that although subjectivity can be suppressed, it cannot be totally wiped out. Every instance of observation is the product of interaction of the mind with the outside world. Without the participation of the mind, there would be no data to start with. Thus, science, however objective, cannot be totally detached from the human mind.[13] Moreover, when physicists deal with the very small scale of the quantum world, mind once more enters into the picture in glaring presence. In at least one interpretation (the Copenhagen version), the outcome of a quantum event does not happen until the observer enters into the scene — taking the measurement.[14]

14.3 Navigating Across Layers of Realities

Physical reality comes in multiple levels. On the lowest level is the quantum world of electrons and subatomic particles. Above that is the chemistry level of atoms and molecules, then the biological world of cells and organisms, then the geophysical world of mountains, oceans and the Earth, and lastly the astronomical world of planetary systems, galaxies, and the universe. At one end is the smallest possible distance — the Planck length, measuring 1×10^{-33} cm.[15] At the other is the diameter of the visible universe, measuring 1×10^{29} cm across. In between and situated "conveniently" near the middle is the familiar world of men, approximately from 1×10^{-5} cm to 1×10^{5} cm, within which everything seems natural, reasonable, and manageable. However, if we traverse across different levels of reality, what is real on one plane becomes irrelevant and even nonsensical on another. For instance, quantum physics defies the common sense of Newtonian mechanics. A paradox can appear at the interface, as in Schrödinger's proverbial cat, which is said to be both alive and dead in the quantum state until brought to the Newtonian world.[16] One can never be attracted to the opposite sex by analyzing the molecular composition of the body. Compassion, fear and other emotions reside in higher animals but are absent in microbes. Religion may appear delusory from the evolutionary stand point, as may romantic love from thermodynamics. On the other extreme, the cosmic scale of time makes human life seem too short to be worth living at all.

14.4 *Self* and Existential Anxiety

Endowed with the faculty of introspection, each person at one point in life realizes, rather shockingly, his or her own mortality, leading to an existential crisis (Fig. 14.2). This painful moment of truth is the rite of passage that propels us to seek refuge and to anchor our ephemeral existence on something more permanent, preferably eternal. Paul Tillich said that the finitude of life is the origin of religion.[17] Earlier, Augustine, the Bishop of Hippo, appealed to God in the opening remark of his *Confessions*, "Our heart is restless until it rests in You." Nonetheless, not everyone finds refuge in a hypothetical, anthropomorphic God. The proofs of God's existence given by Thomas Aquinas are no longer as

Fig. 14.2. The existential crisis. [Permission Patrick Hardin.]

persuasive as they were. (See, for example, Bertrand Russell's *Why I Am Not a Christian*.)[18] Hell as the deepest core of Earth where sinners live in eternal flame, and Heaven as the uppermost stratum of sky full of eternal bliss, and the unfortunate misstep of our first ancestors resulting in mankind's original sin, have all but lost their metaphorical appeal. Since the nineteenth century, existential philosophers discarded the notion of an omnipotent, personal God as the underpinning of all beings. However, stripping all external support and exposing each personal *self* to a cold and bottomless void does not alleviate the existential predicament.

Nevertheless, there are alternative ways to seek man's place in the universe, ways to find solace to man's existential woe, and that is to fuse one's *self* with the natural world — the universe. Here is an excerpt from the book Laozi (*Lao–Tzu*) on the ultimate reality[19]

> "There is this Thing, undefined and pre-existing Heaven and Earth, silent, solitary and unchanging. It goes on forever, endlessly. It is the mother of the world. Not knowing its name, I dub it Dao (Way); compelled to name it, I call it The Greatness…Man conforms to Earth; Earth conforms to Heaven; Heaven conforms to Dao; Dao conforms to Nature."[20]

Dao (Tao) is the mysterious and unexplainable ground of being underlying all things. This naturalistic worldview of Daoist (Taoist) philosophy predates Spinoza by two thousand years.

Spinoza used the word "God" to signify a concept that was different from that of traditional Judeo-Christian-Islamic monotheism. Spinoza's cool, indifferent "God" is the antithesis to the concept of an anthropomorphic, fatherly God who cares about every aspect of humanity. Following is a description of Spinoza's God:

> "This single, eternal, infinite, *self*-caused and necessary principle of things is called God or Nature. God is not, as Descartes held, apart from the world an external transcendent cause acting on it from without (theism), but in the world, the immanent principle of the universe. God is in the world and the world in him. He is the source of everything that is (pantheism). God and the world are one. Cause and effect are not distinct here; God does not create in the sense of producing something separate from, and external to himself, something that can exist apart from him; he is the permanent substance, or substratum or essence in all things... God is the universe conceived as an eternal and necessary unity, an organic whole, a unity in diversity. Spinoza expressly denies personality and consciousness to God. He has neither intelligence, feeling, nor will; he does not act according to purpose, but everything follows necessarily from his nature, according to law; his action is causal, not purposive."[21]

Albert Einstein concurred with Spinoza. "I believe in Spinoza's God who reveals himself in the orderly harmony of what exists, not in a God who concerns himself with fates and actions of human beings." These words were given by Einstein, upon being asked if he believed in God by Rabbi Herbert Goldstein of the Institutional Synagogue, New York, April 24, 1921, published in the New York Times, April 25, 1929.[22] Einstein suggested that behind anything that can be experienced there is something that our minds might never grasp, whose beauty and sublimity reach us only indirectly. "This is religiosity," he said, "a cosmic religious feeling that stays above and beyond the ordinary religious sense of fear and morality."[23]

In Indian thought originated 2,500 years ago, the personal *self* is to be identified with the omnipresent, all-comprehending eternal *self* — "Aum," better known as Brahman.[24,25] The Greek philosopher Plotinus (204–270 C.E.) taught that there is a supreme, transcendent "One," infinite, indivisible and indistinguishable universe. In China, the strife to unite with the universe has been the aspiration of intellectuals for thousands of years. Freud noted (attributing to Romain Rolland) that the sense of boundlessness and oneness with the world is a natural human feeling. It is not an invention of any organized religion, he said, but rather a source of it, and in fact it has frequently been incorporated into their belief systems.[26]

To sum up, the attainment of a cosmic *self* is the highest form of existence. Whether or not one believes in a personal God is a matter of taste and, for me, of secondary importance. But that *Oneness*, that cosmic *Wholeness*, is that which we all seek to be with.

What, then, is the meaning of *self*, of life, and of being? To understand this question, we must analyze what we mean by "meaning." Meaning refers to the connection between parts; explanation relates one thing to another; and value is the worth of one thing against another. All are relative entities arising from the presence of parts or fragments. In an undivided Oneness, all parts vanish and so does meaning (Fig. 14.3). In our personal life, self undergoes metamorphosis. When born, *self* starts with no explicit meaning. As a person matures, multiple meanings accrue — career goals, aspirations, desire, competition, success or failure, fulfillment or despair. In the end, all meanings return to the starting line, and life resumes the original purity — simple and natural. The merging of the personal *self* with the cosmos results in the transcendental *self*, which is devoid of meaning. It is not that meaning is lost; it is that the Wholeness leaves no room for meaning to surface.[27]

If I am compelled to add more, I would say that, at least for some people, the "meaning" of life lies in the quest of its meaning. There are

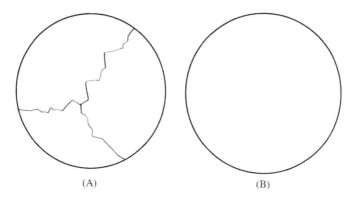

Fig. 14.3. The meaning of meaning. (A) Meaningfulness; (B) Meaninglessness.

concrete examples. Turn around and look at the Pyramids of Giza, the Taj Mahal of India, the Angkor Wat of Cambodia, the Borobudur of Java, and the countless magnificent mosques, temples and cathedrals studding every corner of the continents. Are they not erected for the search of meaning? In all probability, the ultimate answer will never come, but just the same, the search will continue — till the end of time. It is through this endeavor that mankind elevated itself from the level of the animals. It is in this manner that culture and civilization flourished.

The perennial question of existence is best portrayed in Charles Ives' unsettlingly beautiful music score, *The Unanswered Question* (composed in 1906), in which a continuous contemplative mood in the background is punctuated seven times by an impatiently inquisitive trumpet, only to be returned to eternal silence and solitude after a futile and frantic search for an ultimate answer. The question, Ives seems to say, is better than the answer.[28]

14.5 What is Religion?

Religion is the question raised when our transitory *self* confronts the immensity of the universe. It is an endless search without a clear-cut answer. Nonetheless, while pursuing this, a union is forged between the finite and the infinite, the fleeting and the eternal, and the process elevates *self* to the highest level of being. This union, this zeal, is at the heart of religiosity.

Religion, in its broadest sense, is an antidote to existential anxiety.[29] Without it, man cannot live at ease with himself. Over the course of history, religion takes on many forms and a great variety of rites, each suited for the stage of social development and cultural attainment, and of course colored by ethnic idiosyncrasies. The tribal idolatry, the fear of Hell and the lure of Heaven, the plea for a Savior, the strife to escape from the tribulations and travails of endless reincarnations — and, last but not least, for a select few, pure philosophical contemplation — all testify to the need of religion one way or another.

Perhaps a note is useful to explain how philosophical contemplation helps in arriving at religiosity. Here is an example. Einstein confided that it is difficult to explain a cosmic religious feeling to someone who is entirely without it, yet this same feeling is the strongest and noblest incitement to scientific research. A serious scientist, in his opinion, is profoundly religious.[30]

In the last analysis, all the multiple forms of religion are but diverse routes to a common end: the melding of the personal *self* with the universe. If there is an ultimate aim of life, this must be it.[31]

14.6 An Aside: Origins of Religion in Human History

The origin of religion is the origin of existential predicament. Thinking beyond the visible and tangible is a prerequisite to religious beliefs. Following are evidence of religion in prehistoric times[32]:

— 100,000 years ago: Blombos Cave in South Africa — geometric designs.
— 95,000 years ago: Qafzeh in Israel — deliberate burials.
— 65,000 years ago: Burial practices of Neanderthals.
— 30,000 years ago: Upper Paleolithic time; France's Grotte Chauvet — animal cave paintings; Germany's ivory carving of

female figurine ("Venus of Hohles Fels") and half-human, half-animal figurines ("lion-men").
— 11,000 years ago: Turkey's Gobekli Tepe — first temple (rows of standing stones of animal figures).
— (? years ago): Turkey's Catalhoyuk with evidence of religious rite (feasts with wild bulls and burials of ancestors beneath houses).
— 5,000 years ago: Mesopotamia and Egypt — organized religious hierarchies.

14.7 The Universe Inside Us

Our physical body speaks a lot of the universe, way back into billions of years. Within the confine of our skin are common elements, among which oxygen, carbon, hydrogen, nitrogen, calcium and phosphorus make up 99% of our body mass. Most of these were churned out in the furnace of stars from primordial hydrogen and helium. In addition, our body also contains traces of heavy elements that were forged only at the moment of supernova explosions, violent episodes in the countless galactic birth-and-death cycles. Metaphorically, we are orphans of distant stars and foundlings of supernovas. Every cell inside us bears signatures of a long cosmic lineage.[33]

Here on our planet, chemical elements circulate through generations of living things since the congealment of Earth. If atoms had memories and electrons could keep tabs on their former lives, my physical body could easily be the outcome of serial reincarnations of trilobites, coelacanths, pterodactyls, and tyrannosaurs; and the oxygen I breathe in at this moment could have passed through the lungs of Ramses II, Hammurabi, Socrates, Confucius, Alexander, Genghis Khan, Peter the Great, and countless other commoners and noblemen, before lodging by chance in my body. The great recycling continues well into the end of time.

Indeed, we do not own the molecules inside us. They are part-us, part-universe. The atoms of our physical *self* threaded through the cosmos long before our birth, and will continue doing so long after we are gone.

14.8 Mortality as a Source of Creative Drive

Life stands out only when it is juxtaposed against death. Death is objectively inevitable yet subjectively impossible, as absence of consciousness is beyond the realm of our mental reality. Like the 100 billion people who trudged through the Earth before us, we will all slip into the hinterland of nothingness. Among all animals the recognition of mortality and the attempt to defy it is perhaps unique to mankind.

Humans formulate the immortality of a non-tangible *self* (soul or its equivalent) as a way out of this ephemeral life, but the independent existence of "soul" apart from the physical *self* lacks credible evidence. Humans also seek physical immortality by taking potions or doing special mental/physical exercises, although no one has ever succeeded in living forever. Nor has the resurrection of the physical *self* ever been verified. Nonetheless, humans do succeed in building legacies as reminders of themselves after their passing. As Ben Franklin wrote in his mock epitaph, "The body of B. Franklin… lies here, food for worms. But the work shall not be wholly lost:for it will, as he believed, appear once more, in a new and more perfect edition, corrected and amended by the author." Others build empires, write poetry, compose music, paint murals, and make discoveries — things that they hope can outlast them after they are gone. The fear and defiance of death and the metaphorical preservation of a *token self* is a driving force that helps build civilization. How amazing that, through legacies in writings and the creative arts, humans can communicate with their remote ancestors, and dialogue with their distant descendants. The cross-generational flow of ideas and feelings never occurs in other species.

14.9 When Mountains are Mountains Again

Einstein said that the most beautiful emotion we can ever experience is the mysterious. Alfred North Whitehead expressed the same sentiment: "Philosophy begins in wonder. And, at the end, when philosophic thought has done its best, the wonder remains." However much progress we will make in the future, something may remain unattainable, and this should give us a sense of humility, a stark reminder that, after all, we are just infinitesimals in a vast universe.

Here is a teaching from a Chan (Zen) master of the Song dynasty, who talked about the three stages of epistemic attainment: "At the beginning, I saw mountains as mountains, and rivers as rivers; half way through, I did not see mountains as mountains nor rivers as rivers; but at the end of my practice, I see once more mountains as mountains, and rivers as rivers."[34] When we first face reality, we know it intuitively and see the whole (mountain). Once we acquire knowledge and become analytic, we see the details (grains of sand — or, down to the molecules, atoms and quarks) and lose the whole. But when we regain the intuitive perspective, we see the whole as well as the details. Thus the mountain is again a mountain, not just a collection of sand particles or silicon dioxide. The analogy is like seeing the forest as well as the trees. The second wholeness is the fruit of wisdom that transcends the initial wholeness of ignorance and the mid-stage fragmentation of knowledge.

Everything is as it is, and this alone is most beautiful and wonderful. *Being* as such is a mystery, inscrutable, defying explanation and understanding. Amidst this cosmic vastness and endlessness are our minuscule *selves*, fragile and impermanent. It is only by linking the minuscule with the immense, the fleeting with the constant, that we instantiate the beauty and wonder of our existence. In this manner, *self*, which initially "crystallizes" out of the world, in the end yearns to return to it.

Notes and References

1. *Zhuangzi* 《庄子》, Chapter 2.
2. Margenau H. (1950) *The Nature of Physical Reality*. McGraw-Hill, New York; Kuhn TS. (1970) *The Structure of Scientific Revolutions*. (Foundations of the Unity of Science, Vol. II, No. 2), 2nd ed. Univ. of Chicago Press.
3. Only 4% of matter in the universe is visible by their electromagnetic wave emission (normal matter); 25% is not visible but is only gravitationally detected (dark matter); 70% is undetectable energy (dark energy).

4. See, for example, the following statement from Heisenberg: "The existing scientific concepts cover always a very limited part of reality, and the other part that has not yet been understood is infinite." See: Heisenberg W. (1958) *Physics and Philosophy*. Harper & Row, New York, Chapter XI.
5. For an example of the unknowable, see the opening sentence of *Dao De Jing* (*Tao Te Ching*) 《道德经》 by Laozi (Lao Tzu, 老子): "The truth that can be stated (in human language) is not the eternal truth." Youlan Feng (冯友兰) suggested that we have to know a lot in order to better appreciate the unknowable. To illustrate this point, he retold an ancient Chan (Zen) story, which goes as follows. There was a Chan master who would raise his thumb in silence whenever he was asked about Buddhism. One of his disciples imitated him. Upon seeing this, the master quickly chopped off the boy's thumb. The boy cried out in pain and ran away. The master summoned the boy back and showed his own thumb. Thereupon the boy attained sudden enlightenment. See: Yu-lan Fung (1948): *A Short History of Chinese Philosophy*, Macmillan, New York; Free Press, New York, 1966, Chapter 28. [Note: Yu-lan Fung is the old Wade-Giles romanization of the Chinese characters of Youlan Feng.] This perplexing story is subject to different interpretations; that of Feng is that we must amass positive knowledge in order to appreciate its negative side, just like we have to know what "two" means before knowing what "negative two" stands for. The bottom line is, in the pursuit of science, the endpoint is the unknown (but presumably knowable), whereas in the pursuit of philosophy, the climax is the unknowable, or the mysterious.
6. Churchland PS. (1986) *Neurophilosophy*. MIT Press, Cambridge, MA; Dennet D. (1991) *Consciousness Explained*. Little, Brown & Co. Boston; Churchland PM. (1995) *The Engine of Reason, the Seat of the Soul*. MIT Press, Cambridge, MA.
7. Rees MJ. (2000) Just Six Numbers. Basic Books, New York.
8. In an attempt to unite the four fundamental forces of nature, the String Theory was proposed but it does not provide a unique answer. It generates an astronomical number of 10^{500} solutions, to be represented by 10^{500} universes, a figure most scientists find disturbing. See: Susskind L. (2006) *The Cosmic Landscape: String Theory and the Illusion of Intelligent Design*. Back Bay Books.
9. Another theory, the Loop Quantum Gravity, denies any fundamental building blocks in the universe. Space and time and elemental particles come

out of a network of abstract relationships, which are not really objects. Both the string theory and the loop theory are currently not verifiable by experimentation. See: Smolin L. (2006) *The Trouble with Physics: The Rise of String Theory, the Fall of a Science, and What Comes Next*. Houghton Mifflin, Boston; Smolin L. (2001) Three Roads to Quantum Gravity. Basic Books, New York.
10. Penrose R.(2005) *The Road to Reality*. Knopf, New York.
11. Barrow JD. (1992) Shaking the foundations of mathematics. *New Scientist*. Nov. 21, 1992.
12. Dehaene S. (1997) The Number Sense. Oxford Univ. Press, New York.
13. Polanyi M. (1957) Scientific outlook: Its sickness and cure. *Science* **125:** 480–484; Polanyi M. (1969) Objectivity in science — A dangerous illusion? *Scientific Res April* 28, 1969, pp. 24–26.
14. Quantum mechanics does not yield a definite description of an objective reality but deals only with probabilities. The act of measurement, i.e., participation of mind, causes the set of probabilities to randomly assume only one of the possible values.
15. Planck length is the smallest possible distance that still makes sense to physics.
16. A principle in quantum theory called "superposition" says that an object is in all possible states as long as you do not check, but once you look into it, only one appears. Erwin Schrödinger in 1935 proposed the following thought experiment. Put a cat in a steel box along with a vial of hydrocyanic acid and a radioactive substance. The experiment is set up so that when one atom of this substance decays during the test period, it will trigger a sequence of events that will break the vial and kill the cat. As long as the box stays closed, you will not know whether this has happened, and the cat is said to be both alive and dead, or neither alive nor dead. But once you open the box and look, the superposition ceases ("collapse" of the states) and the cat will be found either alive or dead in reality. Superposition occurs only in the subatomic scale such as the electrons, but it becomes a paradox in the human world.
17. What he actually wrote was, "God is the answer to the question implied in man's finitude." See: Church FF. ed. (1987) *The Essential Tillich*. Univ. of Chicago Press, Chicago, Chap. 1, p. 11.

18. Russell B. (1927) Why I am Not a Christian. (A lecture given in 1927), in: Edwards P. ed.(1957) *Why I am Not a Christian.* Simon & Schuster, New York.
19. Laozi: *Dao De Jing* 《老子道德经》, "The Scripture of Dao and De" by Laozi, Chap. 25.
20. Translated from the Chinese text by Ramon Lim. Original in Chinese: 有物混成,先天地生。寂兮寥兮,独立不改,周行而不殆。可以为天下母。吾不知其名,字之曰"道";强为之名,曰 "大"。… 人法地,地法天,天法道,道法自然。
21. Thilly F. (1914) *A History of Philosophy.* Henry Holt, New York, Chap. 47, pp. 296 & 307.
22. Clark RW. (1971) *Einstein: The Life and Times.* World Publishing, New York, p. 413; also cited as a telegram to a Jewish newspaper, 1929, Einstein Archive 33-272, from Calaprice A. ed. (2000) *The Expanded Quotable Einstein.* Princeton Univ. Press, Princeton, NJ.
23. Einstein A. (1949) The World as I See It. Trans. by Harris A. Philosophical Library, New York, pp. 25 & 26 (original in 1934).
24. The Upanisads.
25. The Indian thought of cosmic self has a hint of an all-pervading consciousness whereas the Daoist thought of China is totally natural and impersonal, but this of course is subject to interpretation.
26. Freud S. (1930) *Civilization and its Discontents.* Hogarth Press, London; English trans. by Riviere J.
27. Another way to say it is that being has no intrinsic meaning. Its meaning arises only when it is taken in reference to something else.
28. Swafford J. (1996) *Charles Ives, A Life with Music.* Norton, New York.
29. A broad definition of religion does not necessarily include a personal God.
30. Einstein A. (1949) *The World as I See It.* Trans. by Harris A. Philosophical Library, New York, pp. 26 & 28 (original in 1934).
31. It is interesting to note that Western monotheistic religions emphasize the assertion of self, as in the resurrection of the physical body and its ascension to Heaven; in contrast, Buddhism and Hinduism stress the denunciation of the biological self as the road to liberation.
32. Culotta E. (2009) News focus: On the origin of religion. *Science* **326:** 784–787.

33. All elements in the universe except hydrogen are spontaneous fusion products of the stars, but this process, which is exothermic (giving out energy), goes up to the formation of iron only; elements heavier than iron need energy input (endothermic) to form, which can occur only in a supernova explosion.

34. This famous statement is a Chan classic attributed to Qingyuan Weixin (青原惟信) of Song dynasty (not Tang as reported by some authors). From: *Jiatai Pudeng lu* 《嘉泰普燈录》 CBETA, X79, No. 1559, p. 327, a24-b4. The original Chinese text is: "老僧三十年前未参禅时, 见山是山, 见水是水。及至后来, 亲见知识, 有个入处, 见山不是山, 见水不是水。而今得个休歇处, 依前见山只是山, 见水只是水。" The literal translation is: "Before I had studied Chan for thirty years, I saw mountains as mountains, and waters as waters. When I arrived at a more intimate knowledge, I came to the point where I saw that mountains are not mountains, and waters are not waters. But now that I have got its very substance I am at rest. For it's just that I see mountains once again as mountains, and rivers once again as rivers." See: Watts A. (1989) The Way of Zen, Vintage Books, New York, p. 126. This statement appears in slightly different forms in D.T. Suzuki's writings. See Suzuki DT. (1949) *Essays in Zen Buddhism*. First Series, Rider & Co. London, 1926, p. 22; Grove Press, New York, 1949, p. 24. Also see: Suzuki DT. (1955) *Studies in Zen*. Dell Publishing Co. New York, p. 187.

Chapter 15 Epilogue: And the Quest Goes On

Ever since the Paleolithic hunters set their eyes in the starry sky, humans started to wonder about their place in the universe. Over thousands of years we have acquired enormous amounts of knowledge, rationally and intuitively. But each layer of understanding reveals a deeper layer of mystery, and the quest goes on without end. The destination seems forever inaccessible, but the journey itself is where the meaning lies.

Appendix A: Neurotransmitter Structures

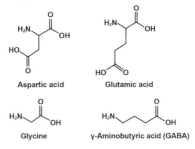

Appendix B: Organization of the Nervous System
(Showing structures mentioned in this book)

I. **Cerebrum (cerebral hemispheres):**

A. **Outer layer (cortex):**
1. **Frontal cortex:**
Prefrontal: ventromedial, dorsal lateral
Primary motor, pre-motor, supplementary motor
Motor speech center
(Anterior cingulate)*
(Insula)*

2. **Temporal cortex:**
Hippocampus: dentate gyrus, CA3, CA1, subiculum
Entorhinal cortex
Parahippocampal cortex
Auditory area, auditory speech center

3. **Parietal cortex:**
Somatic sensory area
Association areas
(Posterior cingulate)*
(Insula)*

4. **Occipital cortex:**
Visual area

B. **Deep brain structures:**
Nucleus accumbens (in basal forebrain)
Amygdala (in temporal lobe)

Ventral pallidum
Thalamus
Hypothalamus

II. Brain stem: (downward extension of cerebrum)

A. Midbrain:
Ventral tegmental area (VTA)
Periaqueductal gray
Reticular activating system (diffuse)

B. Pons:
Parabrachial nucleus
Locus ceruleus
Raphe nuclei
Reticular activating system (diffuse)

C. Medulla:
Nucleus tractus solitarius
Respiratory center
Cardiovascular center
Reticular activating system

III. Cerebellum: (attached to pons)

IV. Spinal cord: (downward extension of medulla)
Spinal nerves: sensory and motor
Autonomic output: sympathetic and parasympathetic

*Insula is the in-folded region between the fronto-parietal and temporal cortices. Cingulate is the midline extension of the frontal and parietal cortices.

Appendix C: Relative Anatomical Positions

- **Afferent:** going toward the brain.
- **Anterior:** toward the front end.
- **Caudal:** toward the tail end.
- **Central:** in or near the center; central nervous system comprises the brain and spinal cord.
- **Coronal section:** a vertical plane perpendicular to the midline, cutting the brain into front and back.
- **Dorsal:** toward the back part of the body; in human brain, roughly equivalent to "superior."
- **Efferent:** going outward from the brain.
- **Frontal:** toward the front end.
- **Inferior:** below or lower down.
- **Lateral:** away from midline toward the sides.
- **Medial:** near or toward the midline.
- **Para:** nearby.
- **Peripheral:** away from the center; peripheral nervous system refers to the nerves coming out of the brain and spinal cord.
- **Posterior:** toward the back.
- **Rostral:** toward the head or front end.
- **Sagittal section:** a vertical plane parallel to the midline, cutting the brain into left and right.
- **Superior:** above or higher up.
- **Ventral:** toward the belly; in human brain, roughly equivalent to "inferior."

Appendix D: Approaches to Explore the Brain

Morphological studies

Gross dissection of the brain provided classical information on brain parts.

Histology: Light microscopic examination of brain sections coupled with various staining procedures provides basic information at the cellular level. A classic example consists of the Golgi staining method (based on silver), which provided crucial information on neuronal connections, giving rise to the "neuron doctrine," the cornerstone of neuroscience.

Histo-chemistry: Microscopic examination of brain sections or cultured brain cells using enzyme reagents, antibodies, polynucleotide probes, coupled with radioactivity or fluorescent reagents, provides chemical information on the cellular and subcellular makeup of brain tissue.

Confocal microscopy provides information on various depths of the brain tissue. It is particularly powerful when combined with fluorescence microscopy.

Transmission electron microscopy (TEM), or simply electron microscopy (EM), visualizes subcellular structures down to the macromolecular level, including the organization of the synaptic junction and the presence of synaptic vesicles that contain neurotransmitters.

Scanning electron microscopy (SEM) reveals surface structures of isolated brain cells or serial brain sections.

Chemical/pharmacological studies

Chemical analysis identifies the molecules (neurotransmitters) responsible for message transmission across the synapse. Molecules that affect neurotransmitters (neuromodulators) are likewise identified. Chemical makeup of neurons and glia (supporting cells in the brain) and their metabolism and intracellular signal transduction, along with gene regulation, are all obtained by chemical studies.

Injection of chemicals into the blood stream or cerebrospinal fluid helps with the discovery of neuro-active drugs.

Micro-injection of chemicals into specific regions of the mouse brain helps to discover drugs affecting the nervous system. Micro-dialysis collects small molecules secreted from specific brain regions, providing a window into their metabolism and helping to identify neurotransmitters and neuromodulators.

Lesion and ablation studies

Surgical removal of brain parts or severance of fiber tracts, coupled with observation of the physiological and behavioral consequences, helps with functional delineation of brain areas. For example, removal of the hippocampus in a patient led to the identification of its pivotal role in memory storage.

Disease states such as stroke, tumor, and developmental deficiency provide a natural setting for the study of function (actually loss of function) pertaining to the defective part of the human brain. For example, observation of stroke patients who have lost the ability to speak led to the identification of the motor speech center in the frontal lobe.

Transient inhibition of function can be achieved in humans by focal cooling of a brain area while awake and observing the deficit in function, such as speech.

Electrical recordings

Electro-encephalography (EEG) is a procedure that records electrical activity of the brain using multiple electrodes placed on the head (scalp). It is a noninvasive procedure as it does not require opening of the skull. It is through EEG that the state of consciousness is conveniently recorded. Clinically, it is very useful to detect seizure activity in the brain.

Electro-corticography (ECoG), also known as intracranial EEG, is the recording of electrical activity from the exposed surface of the brain, using multiple electrodes to detect brain areas responsible for seizure activity, or to detect cortical activity following sensory stimulation (as in visual cortex with respect to retinal light stimulation). The same electrodes can be used for stimulation in order to study the behavioral role of a given cortical area. It is an invasive procedure that is carried out only in patients during brain surgery (for epilepsy) or in non-human animals.

Magneto-encephalography (MEG) consists of recording the magnetic field of the brain surface using a magnetometer place on the surface of the scalp. It has better spatial and temporal resolution than EEG, making it as sensitive as ECoG without having to remove the skull. Its superior time resolution makes it a good complement of functional MRI.

Whole brain **single cell recording** brings animal behavior to the neuronal level, and contributed such important information as the pattern-recognition cells (visual cortex), place- and grid-cells (hippocampus), and the mirror neurons (premotor cortex). Whole brain **multichannel recording** provides information on neuronal interaction and networking.

In vitro single neuron recording provides information on electrical potential across cell membrane: post-synaptic potential; axonal resting and action potentials. Patch clamp (voltage clamp) experiments provide data on ion flux across axonal membranes during neuronal activity.

Electrical stimulation

Deep brain stimulation (DBS) — Stimulation of selected brain areas is used for therapeutic purposes, such as in Parkinson's disease. It has also been used to study the mental and behavioral correlates of a particular brain structure. Historically, electrodes placed in freely moving rats led to the discovery of pleasure centers in the brain.

Transcranial **magnetic** stimulation (TMS) — Application of a magnetic field to the brain through the intact skull has been used for treating depression and anxiety. The procedure is non-invasive. The penetration of the magnetic field is limited to the cortex of the brain and the effect is not as focussed as electrode stimulation.

Transcranial **Direct Current** Stimulation (TDCS) — Delivery of a small current through two electrodes attached to the scalp has been used, attempting to improve cognitive function of patients.

***In vitro* stimulation** coupled with recording of neurons in brain slices can provide information on neuronal "memory" such as long-term potentiation.

Non-invasive imaging

Computerized axial tomography (CAT) — Also called **CT** scan. Imaging of live brain in sections by computer synthesis of multiple X-ray shots of the brain.

Magnetic resonance imaging (MRI) — Imaging of live brain in sections by computer synthesis of multiple nuclear magnetic resonance (NMR) images of the brain. This uses a strong magnetic field in combination with pulses of radio frequency waves, giving detailed images of brain structures without harmful X-ray radiation. Particularly useful for human subjects.

Functional MRI (fMRI) — Detects the level of oxygenated hemoglobin (indirectly the oxygen level), which corresponds to the blood flow of a given region of the brain. It is an indirect measure of metabolic activity (reflecting the function) as opposed to plain MRI, which delineates only the anatomic structures. By far the most useful technique for cognitive neuroscience research.

Diffusion MRI (dMRI) — Detects direction of water diffusion, thus delineating white matter tracts. An exquisite way to map out nerve fiber connections without providing information on activity. Variants of this technology include diffusion tensor imaging (DTI), diffusion spectrum imaging (DSI), and high-angular resolution diffusion imaging (HARDI)

Positron emission tomography (PET) — Maps the metabolic activities of brain regions using a positron emitter such as radioactive fluoro-deoxyglucose, an analogue of glucose that will be taken up by a tissue where glucose is actively utilized. The radioactive tracer has a short half-life and has to be prepared locally in a cyclotron. The positron emitted creates gamma radiation which is then detected by a machine. The resolution is poorer that MRI.

Single photon emission computed tomography (SPECT) — Uses an injected radioisotope to measure blood flow in different areas of the brain. Like PET, SPECT detects functioning of a given area of the brain. They both expose the subject to gamma ray radiation. Compared to PET, the machine for SPECT is more affordable and the radioisotope used is longer lasting and therefore far less expensive. The drawback is that SPECT has poorer spatial resolution.

Genetic manipulations

Genetic modification by **transgenic** and **knock-in/knockout** techniques — In this approach, specific genes can be added or deleted from the genome of an animal. Depending on the promoter used in the added DNA segment, a given protein can be expressed in a particular neuronal type matching the specific promoter sequence, making the approach highly cell-specific. However, the approach is tedious and requires the raising of an animal (mice) colony.

Optogenetic localization and stimulation — Optogenetics applies light technology to genetically modified brain cells. This method, which detects the function of a single cell (or a group of similar cells), in cultures or in the live brain, is of two approaches: (1) Optogenetic *sensing* reveals the activation of a cell by incorporating a reporter fluorescent protein into the genome. A neuron emits light and becomes visible when it is active. (2) Optogenetic *stimulation* consists of incorporating a light sensitive protein into the genome so that the neuron can be switched "on" or "off" by a beam of light into a moving animal through a tiny fiber optic. Depending on the choice of protein that affects the ion channel of a cell membrane, the light sensor can either activate or suppress the activity of a neuron. By engineering the protein to respond to light of a particular wavelength, neurons of different types can be simultaneously manipulated to reveal their interaction. By targeting the foreign gene to a selected neuronal population or part of the nervous system, the method bypasses the tedious transgenic and knock-in/knockout procedures. Its simplicity and incomparably high resolution in time and space surpass those of electrophysiology, making it by far the most promising of all micro-manipulative approaches.

Rapid **gene editing** using **CRISPR** system — CRISPR is a bacterial defense mechanism against viral infection in which a variable RNA target sequence is joined to a constant CRISPR sequence. The target sequence binds to specific target DNA segment while the constant CRISPR RNA sequence recruits an enzyme (nuclease) that cleaves the DNA site. When applied to animal cells, the system permits the specific cutting of a DNA molecule and the introduction of a foreign DNA segment, making possible gene editing with unprecedented speed and ease. This novel approach holds great promise for the study of neuronal functions.

Behavioral studies

Ethology deals with observing animal behavior in their natural habitat, and has led to such interesting findings as imprinting in newly hatched goslings. Animal training is a way to study **classical and operant conditionings**. The latter involves the active participation of the animal under positive or negative reinforcement. Studies of space memory can be carried out in a **Morris water maze** or a **Barnes maze**. Human behavior can be studied by interviewing or by psychological testing. Much information can be obtained by combination with non-invasive imaging procedures.

Computer modeling and informatics

Computer data analysis is an invaluable aid to make sense of the increasingly complex data obtained from neuroscience experimentation. Computer modeling provides a way to test theoretical constructs of brain function.

Glossary

AARS: see Aminoacyl-tRNA synthetase.

Action potential: Membrane potential in the axon that propagates from the axon hillock (the origin of axon) to the synaptic ending (the tip of axon), thus carrying with it the nerve impulse. Action potential is elicited by EPSP (excitatory post-synaptic potential) when the depolarization exceeds a threshold level. An action potential consists of an influx of sodium ion followed by an efflux of potassium ion across the membrane. Once started, an action potential propagates automatically.

Adaptation: Ability of an organism to change in order to survive in a new environment. The theory of evolution provides a mechanism to explain adaptation of a species.

Adaptive immunity: Immunity acquired during the lifetime of an animal.

Adenosine diphosphate (ADP): ATP minus one phosphate. It is a precursor and breakdown product of ATP.

Adenosine triphosphate (ATP): An energy-rich molecule that performs "work" inside the cell, including the synthesis of cellular components and the movement of organelles. It also provides energy for the mobility of cells and organisms, the latter through muscle contraction. Chemically, ATP comprises an adenosine (the nucleobase adenine combined with ribose sugar) attached to three phosphates.

ADP: see Adenosine diphosphate.

Allelopathy: Production of substances toxic to neighboring plants of different species.

Allograft: Transplanting a tissue from one individual to another of the same species.

Amino acid: The building block of a protein, consisting of an amino group, a carboxylic acid group, and a variable side chain, the latter giving distinction to each amino acid. Proteins that occur in nature are made up of twenty different kinds of amino acids, all of the "L" variety.

Aminoacyl-tRNA synthetase (AARS): An enzyme that attaches an amino acid to its corresponding tRNA.

Amnesia: Loss of memory. There are different ways to classify amnesia, such as retrograde versus anterograde in relation to the traumatic event; transient versus permanent with respect to the duration of the amnesia; and focal versus global with respect to the content of loss.

Amygdala: A deep brain structure (gray matter) in the medial temporal lobe. A major emotional center involved in fear, anger, and aggression.

Archaea: Single-celled microorganisms without a nucleus (prokaryotes); similar to but more advance than bacteria; usually thrive in extreme environments.

ATP: see Adenosine triphosphate.

Autograft: Transplanting a tissue within the same individual.

Autonomic nervous system: The part of the nervous system that is not under voluntary control. It regulates the internal organs (heart, lung, stomach, intestine, etc.), along with blood vessels to muscles and glands. It consists of two components, the sympathetic and the parasympathetic.

Axon: A long process extending from the neuronal cell body that conducts nerve impulse (action potential) away from the cell. Each neuron contains only one axon.

B lymphocytes (B cells): Lymphocytes that are capable of producing antibodies.

Base pairing: In nucleic acids, the linking of A to T and G to C through hydrogen bonding.

Blood groups: antigenic properties of red blood cells that can elicit transfusion reaction. Human blood is divided into four categories: A, B, AB, and O.

CaMKII: Calcium-calmodulin kinase II. An enzyme important in the postsynaptic mechanism of learning.

CAT scan: Stands for computerized axial tomography (also known as CT scan). A non-invasive imaging technique providing virtual "slices" of an organ such as the brain by computer processing serial X-ray images taken at different angles.

Central Dogma: A principle in molecular biology stating that the information in DNA is transmitted to another DNA or to a messenger RNA, which in turn is transmitted only to a protein. There are exceptions to this principle.

Cerebral cortex: see Cortex of the brain.

Chaos theory: Also known as deterministic chaos. A non-linear dynamic system whose outcome diverges widely depending on the initial conditions.

Chloroplast: A particle (organelle) in a plant cell that contains chlorophyll and is responsible for photosynthesis.

Chondrites: Meteorites that preserve the original composition during the formation of the solar system.

Cingulate cortex: A part of cerebral cortex in the medial side of the hemisphere. Anterior cingulate cortex is involved in pain interpretation and visceral sensations.

Clonal selection: A mechanism of adaptive immunity by which the activated lymphocytes are allowed to proliferate and expand in number.

Codon: A unit of three nucleotides in a DNA molecule whose sequence codes for a specific amino acid in a protein.

Compatibilist: One who believes that free will is possible in a deterministic world.

Conditioning: A type of associative learning in which an irrelevant stimulus is capable of eliciting a response, when the stimulus was previously paired with (preceding) a relevant stimulus. The term originated from Pavlov's conditioned reflex experiment.

Cortex of the brain: The layer of gray matter forming the outer mantle of the cerebrum (cerebral cortex) or the cerebellum (cerebellar cortex). Cerebral cortex is divided into archicortex, paleocortex, and neocortex, in chronological order of evolution. Neocortex is the highest developed one and has six layers of neurons. In the human brain, the neocortex occupies most of the cerebral cortex and comprises the following: frontal, parietal, occipital and temporal.

CT scan: see CAT scan.

Cybernetics: The study of a system capable of self-regulation, the simplest example of which is the thermostat.

Cyclic AMP: Cyclic adenosine monophosphate. A second messenger important for transducing message from the cell surface to the nucleus.

Defensins: Antimicrobial peptides found in animals as part of innate immunity.

Dendrite: Short processes extending from the neuronal cell body that serve as the receiving end of a synapse. A neuron contains numerous dendrites forming dendritic arborization.

Dendritic cells: Immune cells that present antigenic peptides from pathogens to the T lymphocytes, thereby activating the latter into a killer cell. (Unrelated to dendrites of neurons)

Dendritic spine: Small protrusions on the neuronal dendrites that serve as the post-synaptic component of a synapse. The number of dendritic spines increases with learning.

DNA (Deoxyribonucleic acid): A macromolecule made up of deoxy-ribose sugar, phosphate, and the nucleobases, containing information for inheritance and for the making of proteins.

EEG: see Electro-encephalogram.

Electro-encephalogram: Recording of brain electrical activity through the scalp.

Electron transport chain: A series of steps taking place along the mitochondrial membrane, creating a proton gradient across the membrane. The inward flow of protons at the end of the process provides energy for the synthesis of ATP from ADP.

Entorhinal cortex: The part of the temporal lobe of the brain that provides input and output information to the hippocampus.

Entropy: A measure of disorder derived from the second law of thermodynamics. See "Second law."

Epigenetics: Heritable characteristics of an organism that is not specified in the nucleotide sequence of DNA.

Eusocialism: The instinctive formation of a large social group in insects.

Evolution: Changes in a species across generations. Darwin's theory posits that the underlying mechanism of evolution is the interaction of the heritable variable traits of an organism and the restriction imposed by the environment (natural selection).

Exteroception: Sensation of the stimuli from the outside world, as opposed to those from one's own body.

Functional MRI (fMRI): A method of MRI that detects increase in activity in brain regions through observation of increase in blood flow; useful for correlating brain structures with cognitive function. (See also "magnetic resonance imaging" and "MRI")

Ganglion: In neuro-anatomy, an aggregate of neuronal cell bodies. If located deep inside the brain, it is also called a "nucleus."

Genetic code: The information contained in the nucleotide sequence of DNA that dictates the amino acid sequence of a protein.

Glycolysis: see Glycolytic pathway.

Glycolytic pathway: Metabolic pathway in the cytoplasm that converts glucose to pyruvic acid, before the latter enters the mitochondria.

Gray matter: The outer layer of the brain containing the neuronal cell bodies. So-called because of the grayish coloration. Gray matter makes up the cortex of the brain.

Group selection: A theory of natural selection based on groups of individual organisms whether or not they are genetically closely related.

Gyrus: The surface of the cerebral cortex that forms a ridge between two clefts.

Hebbian theory of learning: A theory that posits that the underlying mechanism of learning involves changes in the neuronal synapse after repeated use.

Hippocampus: An inward extension of the temporal lobe of the brain important for the formation of long-term memory. So called because of its seahorse shape.

Holism: A perspective in which an entity is taken as a whole with its components intricately related. In a holistic view the total is greater than the sum of the individual parts.

Hydrogen bond: The chemical bond responsible for the pairing of nucleobases in a DNA molecule.

Hypothalamo-pituitary-adrenal axis: A chemical channel in which hormone from the hypothalamus stimulates the anterior pituitary gland, which in turn sends hormone to the adrenal cortex (attached to the kidney) to secrete cortisol, a stress hormone with widespread effects on the body and the brain.

Hypothalamus: The part of the brain that is mainly involved in the regulation of the autonomic nervous system and the general well-being of an animal. So called because it is situated below the thalamus.

Innate immunity: Inborn immunity that persists over generations. It is present in plants and animals.

Insular cortex: The in-folding portion of cerebral cortex formed by the temporal and the fronto-parietal lobes. Anterior insular cortex receives sensory fibers from visceral organs.

Internal milieu: Internal environment of a cell and of an organism, which remains constant despite external changes. The ability of an organism to maintain its internal environment is referred to as "homeostasis."

Interoception: Sensation of one's own body components including the internal organs. These sensations are carried through the autonomic afferent fibers to the brain, forming the subjective sense of emotion during an outburst. It also contributes to the perception of one's own bodily self.

Introspection: Perceiving one's own mental activities.

Kin selection: A theory of natural selection based on immediate genetic relatedness.

Leukocytes: Synonymous with white blood cells.

Long-term potentiation (LTP): A physiological phenomenon in which a burst of repetitive stimulation of a neuron leads to the long-lasting facilitation of the synaptic transmission. It is believed to be the mechanism of memory and learning at the neuronal level.

Magnetic resonance imaging (MRI): A non-invasive method to visualize internal organs using the principle of nuclear magnetic resonance. Under strong maganetic field, with the aid of radio-frequency pulses and a computer, detailed images of an organ in a live animal can be constructed. MRI produces more detailed images than X-ray, ultrasound and CT scan do, and is very useful for visualizing brain structures. (see also "functional MRI")

Mega-self: A unit of *self* that is formed by the congregation of smaller *selves*.

Messenger RNA (mRNA): RNA molecule sent from the nucleus to the ribosomes; it contains information for the assemblage of amino acids into protein.

Metabolism: Overall changes of nutrients within a cell as a result of which energy from food is harnessed and utilized.

MHC (Major histocompatibility complex) protein: Found on mammalian cell surface; functions in adaptive immunity and also serves as a marker for recognition of *self* against non-*self* during tissue or organ transplantation.

Mind, theory of: A function of the brain (or mind) to formulate that other persons also have a mind.

Mirror neurons: Neurons that become active when an animal sees another animal performing an act that requires the activity of those neurons. Believed to be related to the mechanism of empathy.

Mitochondria: Small particles (organelles) in the cell in which TCA cycle and oxidative phosphorylation take place.

Motor pathways or tracts: Long fiber tracts that carry message from the brain through the spinal cord to the muscles of the body.

MRI: see Magnetic resonance imaging.

mRNA: see Messenger RNA.

Multilevel selection: A theory of natural selection that includes both kinship and groups of different sizes.

Natural selection: A key mechanism of evolution in which the environment plays a selective role for the survival of species whose inherited characteristics fit the particular environment. In the Darwinian theory, these heritable characteristics tend to vary over generations.

Neocortex: see Cortex of the brain.

Neuromodulators: Any molecule that modifies or regulates the function of a neurotransmitter.

Neuron: Cellular unit of the nervous system that conducts message from one cell to another, through action potential in the axon and neurotransmitters across the synapse between two neurons.

Neurotransmitters: Chemicals of small molecules that carry message across a neuronal synapse.

NMDA receptor: A type of glutamate receptor in the post-synaptic part of a synapse (located in the dendritic spine) believed to be important in the formation of memory and learning, through the mechanism of long-term potentiation (LTP). NMDA stands for N-methyl-D-aspartate.

Nucleic acid: A macromolecule made up of nucleotides linked together with phosphate bonds. It is involved in information transfer and heredity. Nucleic acids are of two types: the ribose (RNA) and the deoxyribose (DNA) varieties. DNA tends to form a double helix whereas RNA assumes irregular shape unique to each molecule.

Nucleobase: Also known simply as the "base." A component of nucleic acid consisting of heterocyclic nitrogenous compounds. It is joined to a ribose sugar molecule to form a nucleoside (in an RNA), or to a deoxyribose to form a deoxyribo-nucleoside (in a DNA). The four nucleobases in DNA are adenine (A), guanine (G), thymine (T), and cytosine (C), with pairing of A-T and G-C. The four nucleobases in RNA are adenine, guanine, uracil, and cytosine.

Nucleoside: A compound consisting of a nucleobase linked to a ribose or deoxyribose sugar.

Nucleotide: A compound consisting of a nucleoside linked to a phosphate group. A nucleic acid such as DNA and RNA is a polymer of nucleotides (polynucleotides).

Nucleus accumbens: A deep brain structure in the temporal lobe involved in motivation and reward.

Nucleus tractus solitarius **(NTS):** A major relay center (in the medulla) for visceral sensation to reach the brain.

Nucleus: In this book, the word is used in two senses. (1) The nucleus inside a cell is the spherical structure containing DNA in the form of chromatin. (2) The nucleus of a brain refers to the aggregate of neuronal cell bodies (gray matter) deep inside the brain.

Orbito-frontal cortex: see Ventro-medial prefrontal cortex.

Oxidative phosphorylation: Phosphorylation of ADP to ATP coupled with the formation of H_2O by the combination of molecular oxygen (from air) and the hydrogen generated at the end of the electron transport chain in the mitochondria.

Parabrachial nucleus: An emotional center located in the pons. Receives sensory fibers from visceral organs via the *nucleus tractus solitarius*, and projects to the higher emotional centers.

Parasympathetic nervous system: The component of the autonomic nervous system that is responsible for counteracting the sympathetic response, i.e., to restore the animal to its resting state.

Peptide: A short version of protein usually containing less than 20 amino acids. Loosely speaking, any polymer of amino acids is a peptide; in this sense a protein is a polypeptide.

Periaqueductal gray: An emotional center in the midbrain. Involved in pain, defense, and sexual behavior.

Phagocytes: Animal cells capable of directly engulfing pathogens. They include neutrophils in the blood and macrophage in the blood and tissues.

Place cells: Hippocampal neurons whose activity correlates with the memory of space.

Plasma membrane: A membrane that encloses a cell. In plants, the plasma membrane is covered by a layer of cell wall.

Post-synaptic potential (PSP): Membrane potential of post-synaptic neuronal cell body. There are two types: the excitatory (EPSP) and the inhibitory (IPSP). The former corresponds to depolarization whereas the latter corresponds to hyperpolarization of the membrane.

Prefrontal cortex (PFC): The part of the frontal lobe cortex minus the areas dedicated to movement and speech, corresponding to the most frontal part of the brain.

Protein conformation: The 3-dimensional shape of a protein molecule.

Protein: A macromolecule made up of amino acids linked together by peptide bonds. It is involved in most of the work in a cell, including metabolism (enzymes), signal transduction, and the building of cellular structures.

Quorum sensing: A way of bacterial communication by which they are able to sense their own population density.

Readiness potential: Action potential detected in the motor cortex before an action is executed.

Red blood cells: Cellular components of blood that contains hemoglobin (imparting a red color) and is responsible for the transport of oxygen.

REM sleep: Rapid-eye-movement sleep. Corresponds to a stage of sleep when dreaming occurs. At this stage the eyes move rapidly while other muscle movements are inhibited. EEG simulates a state of wakefulness.

Reticular activating system: Short afferent fiber tracts that diffusely carry non-discrete sensation (alertness) from the spinal cord to the brain.

Ribose: A five-carbon sugar component in a ribonucleic acid (RNA).

Ribosomal RNA: RNA component of a ribosome.

Ribosome: Small particles in the cytoplasm where proteins are synthesized. It is made up of RNA and protein.

RNA (ribonucleic acid): A type of nucleic acid that differs from DNA in that the deoxyribose is replaced by ribose sugar. RNA does not form a double helical structure but may form base pairing at irregular intervals. RNA is responsible for transfer of message from DNA to protein, and plays a major role in various aspects of protein synthesis and gene expression.

RNA interference (RNAi): A phenomenon in which a short piece (about 21 to 26 nucleotides long) of RNA interferes or modulates cellular function by binding to messenger RNA or to DNA.

Second law of thermodynamics: A natural law that dictates the flow of heat from an object with a higher heat content to one with a lower content. A corollary of the second law is that, in a closed system, physical entities tend to change from order to disorder. A measure of disorder is referred to as "entropy."

Sensory pathways or tracts: Long fiber tracts that carry sensory information from the body to the brain.

Signal transduction: Transmission of message from one part of the cell to another, involving serial protein interactions, frequently incurring phosphorylation and dephosphorylation, leading to the final endpoint of gene activation or repression.

Small interfering RNA (siRNA): Short RNA molecule capable of RNA interference. It is usually referred to a defense mechanism against viral infection.

Somatic: Refers to parts of the body (may include the internal organs).

Steady state: In chemistry, a state in which the concentration of the molecular components remains unchanged despite the presence of constant flux. A steady state differs from one of equilibrium in that the latter ceases to be in flux.

Sulcus: The surface of the cerebral cortex that forms a furrow between two ridges.

Sympathetic nervous system: The component of the autonomic nervous system that is activated by emergency or dangerous situations, resulting in such actions as "fight or flight." It is frequently active in an emotional outburst. The feedback sensation from a sympathetic discharge to the sensory areas of the brain provides a subjective feeling of emotion.

Synapse: A complex structure consisting of a pre-synaptic nerve ending containing neurotransmitters, a post-synaptic component in the downstream neuron containing receptors to the neurotransmitters, and a space (synaptic gap) between the two. The synapse makes possible a one-way chemical transmission of message from one neuron to the next. Synaptic function is subject to regulation and modulation and is responsible for learning.

T lymphocytes (T cells): Lymphocytes that do not produce antibodies. One type of T cell (the killer T cell) destroys cells by direct contact.

TCA cycle: see Tricarboxylic acid cycle.

Terpenoid: A type of volatile organic compound consisting of polyunsaturated hydrocarbons.

Thalamus: A deep brain gray matter structure that serves as the major sensory relay station from the body to the cerebral cortex. It plays an important role in consciousness; also involved in motor regulation.

Toll-like receptor: A type of receptor found on animal cell surface capable of recognizing pathogens; it functions as part of innate immunity.

Transcription: The transfer of information from DNA to messenger RNA.

Transfer RNA (tRNA): RNA molecule responsible for bringing amino acids to the ribosome to be assembled into proteins according to the information contained in the messenger RNA.

Translation: The transfer of information from messenger RNA to protein.

Tricarboxylic acid (TCA) cycle: A series of metabolic steps taking place in the mitochondria by which the acetyl component from food is processed before electron transfer and oxidative phosphorylation.

tRNA: see Transfer RNA.

Ventral pallidum: A deep brain structure in the temporal lobe involved in reward and motivation.

Ventral tegmental area (VTA): A deep brain structure in the midbrain that forms fiber projections collectively called "dopaminergic mesolimbic reward system." Involved in craving behavior, motivation and pleasure.

Ventro-medial prefrontal cortex (vmPFC): The lower (ventral) and midline (medial) part of the prefrontal cortex. It plays a major role in the executive functions of the brain. Also coordinates the various emotional centers, bringing emotion to the conscious level, and serving as liaison between rationality and animal impulse. Also known as orbito-frontal cortex.

Visceral: Refers to the internal organs.

VTA: see Ventral tegmental area.

Watson-Crick model of DNA: A model that visualizes DNA as a twisted ladder (double helix) with a chain of alternating molecules of deoxy-ribose and phosphate on each side, and with the pairing of the nucleobases as the rungs, the pairing being adenine (A) to thymine (T), and guanine (G) to cytosine (C).

White blood cells: Cellular components of blood other than the red blood cells. White blood cells are responsible for immune and other defense functions.

White matter: The deeper part of the brain below the gray matter containing myelinated nerve fibers (axons). So-called because of the whitish appearance produced by the myelin.

Acknowledgments

The author wishes to thank the following:
(alphabetically listed)

- Laird Addis — for reading of the entire book and critical comments on philosophical issues.
- Adel Afifi — for consultation on neuro-anatomical facts.
- Zuhair Ballas — for critical reading of Chapter 6.
- John Donelson — for critical reading of Chapters 1, 2, and 3.
- Kenneth Gayley — for critical reading of Chapters 11 and 13.
- Lourrie Goulet — for generous permission to use her elegant art work for the book cover.
- Adwin Hesseltine — for copyright matters.
- Gregory Landini — for general comments and advice.
- Caroline Lim Starbird — for reading of Chapters 1, 8, 9, 11, 13 and 14.
- Jennifer Lim-Dunham — for reading the entire book and consultation on English usage.
- Victoria Lim — for reading of Chapters 1, 6 and 12.
- Wendell Lim — for critical reading of Chapter 3 and general comments.
- David Noerper — for reading the entire book and useful comments.

- Francesco Orilia — for critical reading of the entire book and useful comments.
- Jon Ringen — for critical reading of Chapters 1 and 11.
- Teresa Ruggle — for assistance with all computer images.
- Morten Schlutter — for consultation on Zen quotations.

About the Author

Ramon (Khe-Siong) Lim (Chinese name 林启祥, born 5 Feb 1933), is Professor Emeritus of Neurology at the University of Iowa. After getting an M.D. degree (cum laude) in the Philippines (1958) and a Ph.D. degree in biochemistry at the University of Pennsylvania (1966), he embarked on post-doctoral work at the University of Michigan, where he participated in research on the molecular basis of animal learning and memory. From 1969 to 1981, he was Assistant and Associate Professors at the University of Chicago, where he engaged in a research program on brain proteins and their effect on the maturation of brain cells. From 1981 till his retirement in 2005, he was Professor of Neurology and Director of the Division of Neurochemistry and Neurobiology at the University of Iowa, where he continued to research on brain proteins and brain cell differentiation, while also engaged in the care of neurologic patients. Lim is a member of seven professional societies related to biochemistry, molecular biology, neurochemistry, neurobiology, neuroscience, and neurology. His research has been funded by the US National Institute of Health and US National Science Foundation, and has resulted in over a hundred original publications in prestigious national and international scientific journals, including *Science* and *Proceedings of the National Academy of Sciences* (USA). In addition, he contributed numerous articles to scientific books. His broad background in chemistry, biology, and basic and clinical neuroscience, along with a life-long interest in philosophical issues, provides him with a unique qualification to simultaneously deal with the two aspects of life (physico-biological and socio-humanistic) in a holistic sense, unified under the common theme of "self." Other than science, Lim is an award-winning artist and a well-known literary writer. His non-scientific creations are collected in the bilingual book *An Anthology of Literary and Artistic Works of Ramon Lim* 《林启祥创作集》 (published in 2008, ISBN 978-971-94113-0-7). While at the University of Iowa, Lim served as an adviser to the world-renowned International Writing Program.

Index

["f"= figures; "t" = tables; "n" = notes and references]

abscisic acid, 104
abulia, 252
acetylcholine, 122, 134t, 174, 185, 337f
action potential, 128–136, 159, 226, 345
adanamide, 190n
adenosine triphosphate (ATP), 18–20, 26, 28, 29, 33, 40
adenosine, 18, 26, 29, 40, 49, 134, 337f
Adolphs, R., 296n
adrenaline (see epinephrine)
aesthetic experience, 192, 201
agnosticism, 66
alexia without agraphia syndrome, 162n
alien hand syndrome, 251
altruism, 65, 202, 261, 265, 266, 273–275, 278, 291
Alzheimer's disease, 201
aminoacyl-tRNA synthetase (AARS), 23, 24, 53-55
aminoacyl-tRNA, 351
amnesia, anterograde, 211, 214
amnesia, focal, 214, 215
amnesia, global, 214, 215
amnesia, retrograde, 214

amnesia, transient global, 215
amoeba, 83, 84
AMPA receptor, 226
amygdala, 143, 144f, 159, 187–192, 194, 197, 199, 200, 219, 339
ant warfare, 287
anterior cingulate cortex, 188f, 189f, 191, 199, 339
antibody, 114, 117–119
anticodon, 22–24
antigen, 115–119, 123
ants, 262–267, 285f, 287
Aplysia californica (sea slug), 220, 221f
Aquinas, T., 322
archaea, 80
Arnold, M.B., 204n
art, as expression of emotion, 200–203
art, as social sentiment, 271, 272
artificial intelligence, 308, 314
Asch, S.E., 271f
aspartic acid, 25, 39, 50, 52, 134, 337f
astrocytes, 135
atheism, 66
Augustine, 298n, 322
autograft, 110

autoimmunity, 122
autonomic nervous system, 144, 184–186
axon hillock, 129f, 131, 132
axon, 128–132

B lymphocytes (B cells), 114, 117, 119
Bacon, F., 317
bacterial quorum sensing, 82, 260
Bard, P., 186, 194
Barnes maze 218, 348
Barrow, J.D., 321
Bartel, D.P., 44
Bataan Death March, 290, 291
Bekoff, M., 279
Benner, S., 14n, 72n
Berger, H., 168
Berridge, K.C., 189f, 205n
Bliss, T., 225
Boltzmann, L., 243, 244
Brahman, 325
brain stem, 142f, 340
brain, as mirror, 158, 309
brain, as projector, 158

Cairn-Smith, A.G., 42, 48
Callahan, M.P., 36
Calvin, M., 49
Cannon-Bard theory of emotion, 186
Cannon, W.B., 66, 186, 194
Capgras syndrome, 200
catastrophe theory, 242
Cech, T.R., 72n
Central Dogma, 24, 25f, 56
cerebellum, 140f, 142f, 340
Chalmers, D.J., 313n
Chan (Zen), 330, 331n, 334n
Changeux, J.-P., 172
chaos theory, 241
chimpanzee territorial war, 287, 288

Chuang-Tzu (Zhuangzi), 299, 311n, 317,
Churchland, P.M., 312n, 331n
Churchland, P.S., 312n, 331n
cingulate cortex, 144f, 188f, 189f, 191, 199
clay, in origin of life, 34, 40–48
clonal selection, 119f
codon, 22–24, 25, 51–55
cogito, ergo sum, 300
comets, in origin of life, 36t
concept cells, 154
conditioning, classical, 217, 222f
conditioning, operant, 217, 218
conflicts, intra-species, 287–292
consciousness, diffuse mode, 170, 173
consciousness, discrete mode, 170, 171
cosmic dust, in origin of life, 36, 38t
cosmic religiosity, 324–327
cosmic self, 315, 325
Crick, F., 25, 43, 58, 172
criminal justice, 283
cyclic AMP, 40, 87, 135, 223f, 226f

Damasio, A.R., 204n
Daoism (Taoism), 324
Darwin, C., 60, 61, 266
Dawkins, R., 4, 5, 63, 278
De Duve, C., 49, 58
Deecke, L., 250
defensin, 113
deism, 66
dendrite, 128–131
dendritic cells, 113–115, 120
dendritic spine, 129f, 130, 132, 145, 224–226, 229
Dennett, D.C., 253n, 302
dentate gyrus, 212, 213f, 228, 339
depolarization, 131f, 132, 226f

Descartes, R., 182, 300, 311n, 324
DNA replication, 26
DNA structure, 21
dopamine, 134t, 188, 193, 194f, 201, 268, 337f
dopaminergic mesolimbic pathway, 189f, 190
drosophila memory, 147, 224

Eccles, J.C., 313n
Edelman, G.M., 172, 313
Eigen, M., 58
Einstein, A. 66, 231, 324, 327, 329
electro-encephalography (EEG), 168-170
electron transport chain, 28, 29f
emergence, 68, 302, 303
emotion, appraisal of, 194, 195
emotion, centers of, 186–192
emotion, definition of, 181, 204n
emotion, expression of, 181
emotion, mechanism of, 196, 197
emotion, sensation of, 192
emotion, sympathetic discharge and, 186, 195, 197
emotions, echelon of, 202t
empathy, 272, 273
endorphin, 160n
enkephalin, 190
entorhinal cortex, 144, 211–213, 227, 228, 339
epigenetics, 64, 277
epinephrine, 185f, 337f
ethylene, 98, 104
eukaryotes, 80
eusociality, 265
excitatory post-synaptic potential (EPSP), 132, 225
exteroception, 164, 192, 196
extremophiles, 80

fantasy, 216, 217
feeling, definition of, 181, 204n
Feng, Y. 319, 331n
Ferris, J.P., 71n
fight-or-flight, 186
flavonoids, 103
formamide, 36t, 38t, 42, 48, 49
Franklin, B. 329
Franks, N. 246
free will, as selector, 245
free will, chances and, 242, 244
free will, definition of, 239
free will, implicit cause of, 238, 240
free will, mind and, 244–246
free will, physiological basis of, 247–249
Freud, S., 157, 207n, 325
frontal lobe, 141f, 142f, 191, 252, 339
Fukushima nuclear disaster, 274

gamma amino-butyric acid (GABA), 134t, 337f
Gauguin, P., 15
Gazzaniga, M.S., 161n
genetic code, origin of, 25t, 51–54
genetic drift, 304
Gilbert, W., 70n
glia, 135
glutamic acid, 25, 35, 39, 52, 134t, 226f, 337f
glycine, 25, 35–37, 39, 50, 52, 134, 337f
glycolysis, 26, 28, 29f
Gould, S.J., 272, 313n
grid cells, 227

habituation, 136, 139, 167, 213, 220–222, 224
Haldane, J.B.S., 265
Hamilton, A., 179
Hamilton, W.D., 265, 266

happiness, what is, 282
Hawking. S., 298n, 318
Hebb, D.O., 144, 220
Heisenberg, W., 331n
Henry, P., 237
Hippocampal formation, 212f
hippocampus, 144f, 209–213, 227–231, 339
histamine, 134, 337f
HM (Molaison, Henry) 210–214
Hobbes, T., 297n
Hofstadter, D.R., 77n
Holmes, O.W., 198
homeostasis, 31
honeybees, 147, 148, 262, 265
Hoyle, F. 9
Hume, D., 198, 240, 243
Hutchinson, C.A., 75n
hyperpolarization, 132, 359
hypothalamus, 144f, 174, 176, 184–187, 191, 192, 196, 197f, 340

imagination, 216, 217
immune surveillance, 115, 124
immunity, adaptive, 116–119
immunity, and self, 119–125
immunity, in animals, 109–125
immunity, in microbes, 81
immunity, in plants, 104–106
immunity, innate, 112–116
incompatibilism, 240
inhibitory post-synaptic potential (IPSP), 132
instinctive behavior, 147–149
insular cortex, 188f, 188–193, 199, 272, 339
intelligent design, 66
interferons, 116
interoception, 164, 185f, 192, 196, 197f
Ives, C., 326

Jablonka, E., 8n, 297n
James-Lang theory of emotion, 183
James, W., 167, 182, 183, 193, 238, 300
jasmonate, 98, 104
Jenner, E., 112
Jennings, H.S, 79, 85, 90, 91, 307
Joyce, G.F., 71n, 72n

Kandel, E.R., 157, 220–223
Kant, I., 32, 231
Kauffman, S., 56–58
Kety, S.S., 127, 312n
Kluver, H., 186, 187
knowledge, sphere of, 318f
Koch, C., 305
Kornhuber, H.H., 250
Koshland, D.E., 13n, 92n
Kuhn, T.S., 66, 330

Landsteiner, K., 110
Langer, S.K., 207n
Laozi (Lao-Tzu), 323
Laplace, P.-S., 241
Libet, B., 250
life, definitions of, 10, 11
limbic system, 143, 144f, 187
Lock, J., 297n
locus ceruleus, 170, 174, 340
Lomo, T., 225
long-term depression (LTD), 225
long-term potentiation (LTP), 225–227
Lorenz, E.N., 241
Lorenz, K., 149, 199, 269
lymphocytes, 109, 114t, 116–120

MacLean, P., 157f
macrophages, 114, 115
magneto-encephalography (MEG), 170, 345

major histocompatibility complex (MHC), 119–121
Margenau, H., 330n
Margulis, L., 293n
materialism, eliminative, 303
Mayr, E., 77n
meaning, 325, 326f
Medawar, P.B., 110, 121
medulla, 141, 142f, 340
mega-self, 258
Mele, A.R., 254n
memory, associative, 213t, 214, 217, 221, 224
memory, declarative, 211, 213t
memory, episodic, 211, 213t
memory, explicit, 211, 213t
memory, implicit, 211, 213t
memory, invertebrates, 219–224
memory, non-associative, 211, 213t, 214, 217, 222, 224
memory, procedural, 213t, 219
memory, semantic, 211, 213t
memory, vertebrates, 224–229
meteorites, in origin of life, 35t
microglia, 135
midbrain, 141, 142f, 143, 340
Mill, J.S., 281
Miller, S.L., 37, 56
Milner, B., 232n
mind, theory of, 164
mirror neurons, 272, 273, 345
mirror test, 164, 165
molecular evolution, 61
monoamine oxidase, 194
Monod, J., 58
montmorillonite, 40, 48
mood, what is, 181
moral code, foundations of, 280–282
morality, definition of, 282, 283
Morowitz, H.J., 49
Morris water maze, 218, 348

Moser, E.I., 234n
Moser, M.B., 234n
motivation center, 189f, 193, 194f, 219
motivation *versus* reward, 190
Murchison meteorite, 35
mutualism, 259
mycorrhizae, 101–103

Nagel, T., 303
Nanjing massacre, 289
Nanking, rape of, 289, 291
neurogenesis, adult, and memory, 228
neurogenesis, adult, in dentate gyrus, 228
neuromodulators, 135, 160n
neuronal workspace model, 172
neurons, 128–135
neurotransmitters, 134, 135, 337f
neutral monism, 314n
neutrophils, 113–115, 117, 274
Newton, I., 231, 322, 318
nitric oxide, 134, 337f
NMDA receptor, 225, 226f
noradrenaline (see norepinephrine)
norepinephrine, 134t, 170t, 185f, 194f, 337f
nucleobases, 19, 20, 21f
nucleoside, 19, 20, 21f
nucleotide, 19, 20, 21f
nucleus accumbens, 188, 189f, 339
nucleus tractus solitarius (NTS), 193, 340

O'Keefe, J., 225
occipital lobe, 142f, 339
Olds, J., 186,188
oligodendroglia, 135
ontology, 309, 314n
orbitofrontal cortex, 191
orexin, 170, 174, 176
Orgel, L.E., 58

original sin, 281, 323
Oró, J., 37
osmotin, 96, 104
oxidative phosphorylation, 28, 29f
oxytocin, 193, 268, 269

pain and pleasure, 190, 198, 199, 312n
pain asymbolia, 199
panpsychism, 306
panspermia, 43
parabrachial nucleus, 189, 192, 193, 340
paradigm shift, 317
parallel dynamics, 153, 172
parasympathetic nervous system, 184
parietal lobe, 142f, 171, 190, 200, 248, 249, 252, 339
Pasteur, L., 16
Pathogenesis-related proteins, 104
Pauli exclusion principle, 17
Pavlov, I., 213, 217
Penfield, W., 154
Penrose, R., 320
peptidyl transferase, 46, 51–54
peri-aqueductal gray, 189f, 192, 340
Perret, M, 11
phantom limb syndrome, 158, 172, 252
pheromones, 148, 194, 264–267, 287, 307
phytoalexins, 103
place cells, 227, 228
Planck, M., 320, 322
plant hormones, 98
plant immunity, 104–106
plant stress hormones, 104
Plotinus, 325
Poincare, H., 241
Polanyi, M., 77n, 332n
pons, 142f, 340

Popper, K.R., 313n
post-synaptic potential (PSP), 134, 345, 359
prefrontal cortex, anatomical location, 141f, 339
prefrontal cortex, echelon of emotions and, 202
prefrontal cortex, evolution of, 141f, 146
prefrontal cortex, memory and, 214t, 230
prefrontal cortex, readiness potential and, 251
prefrontal cortex, superego and, 157
prefrontal cortex, volitional acts and, 247, 248f
prion protein, 64, 65, 224
pro-social hormones, 268, 269
prokaryotes, 80
property dualism, 304
protein conformation, 27f
protein synthesis, schematic of, 22f, 23f
protists, 80, 83

qualia, 176n
quantum mechanics, 332n
quantum superposition, 332n
quorum sensing, 82, 260

Ramon-y-Cajal, S., 91, 130f, 220
raphe nuclei, 170t, 174, 340
rapid-eye-movement (REM) sleep, 169
readiness potential, 250
realities, layers of, 321
reductionism, 67
reentrance dynamics, 172
Rees, M.J., 68, 331n
religion, what is, 315, 326, 327

replicator, 4, 5, 63
response, conditioned, 217
response, unconditioned, 217
resveratrol, 103
reticular activating system, 173f, 340
reward center, 186, 189f, 190, 193, 201, 219
rhizobacteria, 101
Ribo-nucleoprotein World, 46
ribocyte, 45
ribosome, 20f, 22, 23f
ribozyme, 39
Rizzolatti, G., 296n
RNA interference, 64, 104, 116, 359, 360
RNA silencing, 104
RNA World, 39, 41, 45, 46, 49, 55
RNA, messenger (mRNA), 22, 23f
RNA, transfer (tRNA), 22, 23f
robots, 308
Rousseau, J.-J., 297n
Russell, B., 66, 243, 301, 323
Rwandan massacre, 288

Saladino, R., 73n
Salicylic acid, 104
Sally-Anne test, 165–166
Sartre, J.P., 253n
Schrödinger, E., 7, 31, 322
Searle, J.R., 303
selection, group, 65, 257, 266, 278, 288, 291, 292
selection, kin, 65, 266, 278, 288
selection, multilevel, 65, 266, 278
self, as driving force of evolution, 63
self, as unit of natural selection, 65
self, cosmic, 315, 325
self, definition of, 2
self, microbial, 79
self, molecular, 109

self, neurobehavioral, 127
self, plant, 93
self, robotic, 308, 309
selfish gene, 4, 63
sensitization (memory), 213, 220–222
sensory evoked potential, 171
serotonin, 134t, 170t, 174, 193, 194f, 221, 269, 337f
sham rage, 195
Sherrington, C., 91, 192, 220, 300
Singer, T., 272
Skinner, B.F., 213, 218
slime molds, 87–89, 261
Smolin, L., 332n
social glue, 267
solipsism, 306
Sperry, R.W., 145, 155
Spiegelman, S., 61, 62
spinal cord, 140, 141, 144, 185f, 340
Spinoza, B., 324
split-brain, 155, 156
statistical mechanics, 244
steady state, 31, 69n, 360
stentor, 85–87
stimulus, conditioned, 217
stimulus, unconditioned, 217
stream of consciousness, 167, 317
strong artificial intelligence, 308, 314n
superorganism, 8n, 293n
supervenience, 304
Susskind, L., 331n
Sutherland, J.D., 42, 49, 71n
Suzuki, D.T., 334n
symbiosis, 259
sympathetic nervous system 184
synapse, 130–132
synaptic cleft, 130–133
synaptic theory of learning, 220

system, definition of, 7n
Szostak, J.W. 44, 47

T lymphocytes (T cells), 114t, 119–124
Taoism (Daoism), 324
temporal lobe, anatomical location, 142f, 339
temporal lobe, emotion and, 186, 187
temporal lobe, epilepsy and, 175
temporal lobe, memory and, 210–212
termites, 259, 262, 264–267, 275
thalamic pain syndrome, 158, 199
thalamus, 142f, 170–173, 340
theism, 66
theory-of-mind test, 165, 166
thermodynamics, 11, 26, 31, 56, 60
Tillich, P., 322
Time cells, 228
Tinbergen, N., 162n
transcription, schematic of, 22f
transfer RNA, structure of, 22f, 23f
translation, schematic of, 22f, 23f
tricarboxylic acid (TCA) cycle, 28, 29f

Urey, H., 37, 56

value, what is, 315
vasopressin, 193, 268, 269
Venter, C., 75n
ventral pallidum, 188–190, 268, 269, 340
ventral tegmental area (VTA), 188–190, 201, 340
ventromedial prefrontal cortex, anatomical location, 144f, 188f, 191, 339
ventromedial prefrontal cortex, and amygdala, 187
ventromedial prefrontal cortex, and ventral tegmental area, 190
ventromedial prefrontal cortex, as liaison with emotion centers, 189f
ventromedial prefrontal cortex, as part of reward system, 190
ventromedial prefrontal cortex, in aesthetic appreciation, 192, 201
ventromedial prefrontal cortex, in pain and pleasure interpretation, 199
volatile organic compounds (VOC), 97, 98

Wachtershauser, G., 48
Waldrop, M.M., 56
war, as group selection, 291, 292
war, in ants, 287
war, in chimpanzees, 287, 288
war, in humans, 288–292
water paradox, 42
Watts, A., 334n
Wegner, D.M., 251
White, H.B., 70n
Whitehead, A.N., 329
Whitman, W., 93
Wigner, E.P., 299
Wilson, D.S., 266, 278
Wilson, E.O., 8n, 266, 278

Yarus, M., 52, 53

Zayonc, R.B., 198
Zen (Chan), 330, 331n, 334n
Zhuangzi (Chuang-Tzu), 299, 311n, 317